Information Security for Small and Midsized Businesses

Greg Schaffer

Information Security for Small and Midsized Businesses

Third edition published by Second Chance Publishing, LLC.

Editor: Erin Kelley

Cover Design: Christian Storm

Third edition ISBN: 978-1-7330668-6-0

Second Chance Publishing, LLC
PO Box 680551
Franklin, Tennessee 37068-0551
https://www.secondchancebook.org

Introduction

Small and midsized businesses (SMBs) have the same information security concerns and needs as large organizations yet are often hampered by resource limitations. Most large companies have a Chief Information Security Officer (CISO) to lead and manage information security programs, initiatives, and risks. However, the cost of retaining a full-time CISO is often prohibitive for SMBs.

This gap has led to the rise of the virtual CISO, or vCISO role. Sometimes referred to as a fractional CISO, a vCISO is a part-time consultant who works virtually (remotely) as opposed to in the office. Because of the high demand for CISO experience, the virtual nature keeps the cost of engaging a vCISO relatively low. Yet SMBs can dramatically improve their information security risk management posture without a virtual CISO, with the right information and mindset.

This book aims to help get your small or midsized business there. From my firm's numerous client engagements, we have noticed SMBs share many similar information security knowledge gaps. Certain question patterns emerged—"What's the difference between a SOC1 and a SOC2?" "What does a high vulnerability mean?" "How do we know our partners are secure?" "How can we know where our security gaps are?" "What is the best framework to align to?"[1] These are some of the concerns we field while building and managing an effective information security program.

Think of this work, a result of discussing those common needs, as a pocket information security risk management consultant. It is my hope that, by collating and presenting the most significant concerns to SMB information security, this publication can help your SMB begin or continue your journey to a more secure environment. Effective information security risk reduction does not need to be cost prohibitive. Indeed, it works best when integrated with all business processes.

I have striven to present all topics in plain English, focusing more on the business needs, rather than using "giga-mumbo-jumbo" terminology, to paraphrase an early career mentor of mine. That is key, because at its core information security is two things—a business issue, and risk management.

My goal is to provide a simple yet powerful resource to you, the SMB executive, so you may be able to make risk-informed decisions, with or without a virtual CISO. I have organized this publication in a somewhat cohesive order, so subsequent chapters build off the previous, yet each is kept

---

[1]Guidelines, standards, and best practices for ensuring the security of information.

independent enough to allow this to also be used as a reference. In other words, you may choose to read it sequentially or by chapter based on topical interests (or both).

Security is everyone's responsibility. Those are not just words; they are a truth in business that if not heeded can lead to losses and even failure. Don't be that business!

Third Edition Notes

The first edition was an eBook, designed as a lead magnet for our website (enter your email address and download this free publication), and was limited in content. The second edition, released in 2021, greatly expanded the material and was offered as a print edition for the first time.

The feedback on the second edition was enormously positive. For example, at one conference in 2022, a keynote speaker held up the second edition and said that this was one of the best, to-the-point guides regarding information security for small and midsized businesses he had ever read. That sort of feedback, coupled with my passion and calling to help SMBs, prompted me to work on a third edition that is approximately 50 percent larger than the previous.

I have checked all the links in this document for currency and adjusted where needed. I believe a well-referenced document provides value, yet the internet is never static. Therefore, I apologize in advance for any broken links. My suggestion is to search the footnoted site for the referenced article or resource. Often website overhauls keep the same documentation but change URLs, creating dead links.

I will continue to work on subsequent edition updates as time permits and am always looking to improve this work. If you have suggestions for content, please let me know.

About the Author

I entered the virtual CISO realm in 2017 by choice. My calling was to do more with my talents, birthed in the numerous stories of large-company breaches. I think the Equifax breach may have been the tipping point. These huge organizations had the resources and, most of the time, the leadership to prevent breaches, and yet were still compromised. What about the small and midsized businesses?

This was indeed a calling to service, one that I initially resisted.[2] I have held a job constantly since my teens, had never been fired or laid off, and only left a full-time position after securing the next. I craved the comfort and security that only a full-time job could provide. However, growth is often more pronounced when comfort is left behind.

Thus, in 2017, I launched the consulting firm vCISO Services, LLC[3] with the mission of providing quality experience to SMBs to help prevent losses due to information security weaknesses. All our client engagements are led by an experienced CISO, and all our supporting resources have at least five years of experience in the field. With that

---

[2] For a brief look into this walk of faith, see https://youtu.be/6JtVF5ao8Qg

[3] https://vcisoservices.com

baseline, SMBs are confident that they receive quality experience akin to what big corporations employ.

I hope that my over thirty years of experience in information technology and information security will help SMBs improve their information security posture. Indeed, I have expanded efforts to shed light on SMB information security risks and ways to address them, including a weekly podcast (The Virtual CISO Moment), conference presentations, and, of course, this book. This is my focus at this stage in my career. I hope you find value in this book.

Acknowledgements

Information security is a discipline that should not be practiced in a silo. As part of its nature, collaboration on threat intel and best approaches is necessary to arrive at the best approaches to manage risk.

For the third edition, I had the good fortune of working with several volunteer experts in information security. They offered suggestions from typos and grammar to complex issues within information security. Their input made this a substantially better product than I could have accomplished on my own. Specifically, I'd like to thank the following amazing people:

Heather Noggle, who provided both relevant industry knowledge and significant writing advice. I will forever thank, or blame, her for understanding the difference between "less" and "fewer."

Peter Gregory, who not only added valuable content, but also provided author advice, both directly and via his excellent book *The Art of Writing Technical Books: The Tools, Techniques, and Lifestyle of a Published Author*.

Michael Cole, who provided welcome pushback on several sections, particularly the final one regarding virtual CISOs. I most appreciate those who challenge me, as that type of input improves both the product and me.

Tom Rule, whose eye for detail is way beyond anything I could accomplish. Additionally, I felt an immediate "old-school" connection as he provided manuscript feedback in my favorite format: a print out, edited with pen, then scanned.

David Godbee, who is probably the most familiar with not only my industry knowledge but also my writing style and personal life. His suggestions from those perspectives tightened the message and therefore the mission of this work.

Evan Grace, Bill Richardson, and Samson Adekoya, for excellent suggestions that I could not incorporate into this edition but which I will consider for the next one.

Steven Woods and Jerry Mitchell, for putting additional eyes on the manuscript.

Special thanks to my wonderful wife, Vicki, for her constant support and understanding, not only as I have worked on this book, but in all other related endeavors. From the beginning, jumping into the vCISO discipline, she has encouraged me to follow my desire to pursue many paths to help small and midsized business information security.

Finally, and most importantly, none of this would have been possible without the skills and opportunities gifted by God.

Information Technology Security and Information Security

The terms "information technology security" and "information security" are often, and incorrectly, used interchangeably. Understanding the differences is a foundational step in understanding risks to an organization's[4] information.

SANS defines information technology security as "the process of implementing measures and systems designed to securely protect and safeguard information (business and personal data, voice conversations, still images, motion pictures, multimedia presentations, including those not yet conceived) utilizing various forms of technology developed to create, store, use and exchange such information against any unauthorized access, misuse, malfunction, modification, destruction, or improper disclosure, thereby preserving the value, confidentiality, integrity, availability, intended use and its ability to perform their permitted critical functions.[5]

Note the mention of securing "utilizing various forms of technology." Information technology security focuses on

---

[4] Organization, corporate, and business are terms used interchangeably throughout this document.

[5] https://www.sans.org/it-security/

technology methods to reduce the risks to the security of information. To see how this differs from information security, examine its SANS definition: "Information security refers to the *processes and methodologies* which are designed and implemented to protect print, electronic, or any other form of confidential, private, and sensitive information or data from unauthorized access, use, misuse, disclosure, destruction, modification, or disruption.[6]

Notice the broadness of the information security definition, to encompass "processes and methodologies" to protect "print, electronic, or any other form" of information. Information technology security is a subset of information security.

Information security is the work the organization takes to protect the Confidentiality, Integrity, and Availability of information (the CIA Triad).[7] Those corporate security programs that only focus on information technology security are likely not addressing all aspects of the CIA Triad, and incomplete planning and practices can lead to

[6] https://www.sans.org/information-security/

[7] Many often think of information security as solely protecting information confidentiality, but the "information security triad" lists availability and integrity as components of information security as well. See https://www.csoonline.com/article/3519908/the-cia-triad-definition-components-and-examples.html.

information security program gaps that management is never aware of until after a breach.

What are some attributes of information technology security and information security? Information technology security controls[8] may include:

- Firewalls
- Endpoint Detection and Response measures
- Disaster recovery

Whereas information security also addresses:

- Risk assessment
- Vendor management
- Business continuity

What about Cybersecurity? Cybersecurity is a term that has been around for at least twenty years and has taken on different meanings to different people during that period. Originally cybersecurity referred to information technology security, as cyber is defined as "of, relating to, or involving computers or computer networks (such as the Internet)."[9] Merriam-Webster today defines cybersecurity as "measures taken to protect a computer or computer system (as on the internet) against unauthorized access or

---

[8] See the Basic Information Technology Controls section.

[9] https://www.merriam-webster.com/dictionary/cyber

attack."[10] In that sense, cybersecurity is still equivalent to information technology security. However, others interpret cybersecurity to include all aspects of information security. For this reason, I shy away from the term cybersecurity to avoid confusion. Words matter, and when discussing information security, scope must be clear.

Bottom Line: Regardless of terminology, protecting information assets is about managing risk, to include information technology-induced and other risks.

SMB Considerations: Some may still think that since they have a firewall or that they outsource their resources (such as Microsoft 365) that all security issues are addressed. The truth is small organizations have the same information security needs as large ones and should therefore address those with the same mindset and list of concerns. Budgets and toolsets may differ, but the risks do not.

---

[10] https://www.merriam-webster.com/dictionary/cybersecurity

Basic Information Technology Controls

While this book is primarily focused on information security risk management items, the SMB should understand basic information technology security controls. These are technology tools that reduce the risk of loss of information by increasing the defensive posture of the technology environment. The most important categories are summarized below:

- Network Firewall—A network firewall separates two network domains, such as the internet and the corporate network. By applying policies, a network administrator can allow network traffic necessary for business while blocking potentially nefarious communications. A network firewall is a necessary component of information technology security but is not, contrary to some marketing claims, a cure-all. Additionally, if not configured correctly, trust in the network firewall can literally provide a false sense of security.
- Local Firewall—Personal computers may utilize host firewalls that perform similar operations as network firewalls, albeit on a smaller scale.
- Antivirus and Antimalware Software—This software is designed to identify threats that may have reached a computer. This software uses both signatures and heuristics to identify and infer viruses and malware threats so they can be blocked before executing and causing damage.

- Backups—Backups refer to the process of copying data to a separate environment. Backups can restore data that was accidentally deleted, obfuscated by ransomware encryption, lost due to hardware failure, or other reasons. Backups should be tested periodically by restoring test or other files to validate the business continuity plan requirements and restoration functions.
- Intrusion Detection/Prevention System (IDS/IPS)—These systems detect threats to a network and its systems based on behavior. Suspicious traffic may be identified for further analysis (IDS) or blocked, either internally (IPS) or by the IDS communicating with a network firewall.
- Virtual Private Network (VPN)—VPN is a method of secure communication, ensuring the information is encrypted (scrambled) to prevent use if intercepted. Computers may connect via a VPN to corporate resources, or different corporate locations may be connected over the public internet securely using VPN technology.
- Security Information and Event Management (SIEM)—SIEMs processes and analyzes logs from various network components (e.g., servers, routers). The sheer volume of the logs produced makes human analysis impossible. A SIEM analyzes the logs, leveraging correlations between systems, to present only those items that present a significant threat and need to be acted on.

- Data Loss Prevention (DLP)—DLP tools identify traffic flows (including USB transfers) that may contain confidential information, such as Social Security Numbers and either block or quarantine for examination the data stream. DLP is useful both for correcting accidental sending of confidential information and for blocking unauthorized attempts of data exfiltration.
- Encryption Tools—Unlike VPNs which encrypt a communication stream, encryption tools encrypt the data directly. Confidential information both at rest (stored) and in transit must be encrypted. Otherwise, unauthorized access or interception may result in data exfiltration and readability. Encryption is available in many levels and strength generally increases in effectiveness over time, as computing power to break older encrypted communications becomes available through technology progress, necessitating improvements.
- Patch Management Software—Software running on computers, servers, and network infrastructure needs to be up-to-date and "patched" (updated) with revised code to block recently identified vulnerabilities. Given the complexity of software, vulnerabilities are constantly discovered and subsequently patched.
- Network Access Control (NAC)—NAC prevents access to a network unless a predefined set of conditions is met, such as denying access to a

computer attempting to join the network because its software is out-of-date and has known vulnerabilities.

- Authorization and Authentication Verification—Many confuse authorization and authentication, but they serve two distinct purposes (and therefore present different potential risks). Authorization is "the process of giving someone the ability to access a resource."[11] Authentication is "the process of determining whether someone or something is who or what they say they are."[12]
- Endpoint Detection and Response (EDR)—EDR identifies machines that have vulnerabilities in near real time without relying on a network vulnerability scan. EDR generally requires an agent (small piece of software) loaded on endpoints (e.g., laptops computers, servers).
- Extended Detection and Response (XDR)—XDR is an integrated set of tools to "detect, analyze and respond to potential threats in real time, covering the network, cloud, endpoint and email security domains."[13]

---

[11] https://auth0.com/intro-to-iam/what-is-authorization

[12] https://www.techtarget.com/searchsecurity/definition/authentication

[13] https://www.forbes.com/advisor/business/what-is-xdr/

- Web Filtering—This filtering prevents connectivity to known malicious websites that may distribute malware or attempt to catch credentials as well as sites not allowed through corporate policies (e.g., pornography, gambling).
- Email Security Tools—This category includes software such as spam filters to prevent or mitigate risks due to malicious emails. Email is a vector commonly used by criminals to introduce malware or steal credentials.
- File Transfer Tools—Many organizations identify the need to routinely exchange large datafiles with other organizations, and sometimes these files are too large to send by email. A robust file transfer system manages the safe and secure transfer of files to and from outside parties.
- Two-Factor Authentication (2FA)—Two-factor authentication, a subset of multi-factor authentication (MFA), requires something you know, have, or are, in addition to a username and password, to log in to a system. A frequent 2FA control is an authenticator app on smartphones that provides a code needed to login. MFA is a broader term that requires two or more factors to authenticate (e.g., password, authenticator code, fingerprint).
- Vulnerability Scanning—Vulnerability scanning is a process (usually automated) to identify vulnerabilities in systems, often present in

unpatched or misconfigured machines. These scans may be performed on external or internal resources.

- Penetration Testing—Human action to attempt, with prior authorization, to exploit vulnerabilities before a criminal does so the vulnerabilities can be identified and then resolved.

These controls may be classified as follows:

- Preventive—Controls designed to prevent an action from occurring. A network firewall is a preventative control.
- Detective—Controls to determine if there are anomalous actions occurring or have occurred. A SIEM is a detective control.
- Corrective—Controls to ensure that items discovered are prevented from occurring in the future. Backups are a corrective control.

Controls are not always technical and may be administrative. For example, an information security policy is a preventive control, an information security risk assessment is a detective control, and business continuity and disaster recovery plans are corrective controls.

Implemented together, these tools are a component of a layered security-based information security strategic plan.[14] If one layer, or control, fails, another may catch the threat. The goal is to maximize the organization's information security posture to minimize loss of information or services.

Bottom Line: A layered security approach, known as defense in depth, is the most effective in reducing information security threats. Implementation requires a careful understanding of the tools in place, how those tools are managed, and identification of gaps in coverage.

SMB Considerations: Layered security is not just for large organizations. While SMBs generally manage less complex technology and fewer information environments, they should, at the core of their information security strategic and tactical plans, employ and manage a layered information security program.

---

[14] See the Strategic Planning section.

Compliance and Information Security

There is a common saying among information security professionals: *compliance is not security*. This statement reflects a simplified representation of a complex and risky mindset that begins with a false sense of security (pun intended) that just because an auditor has deemed an organization compliant with a particular regulation or standard that they are secure from threats. That mindset is a significant threat to comprehensive information security as it reduces the perceived need to enhance the information security posture when there may be gaps unexplored and unknown to executive management and the board of directors. That said, it is better to at least be compliant, as this will increase information security posture as opposed to doing nothing at all.

There are two driving factors behind this mindset: scope and time. What an audit encompasses may not cover the full spectrum of threats. Take a Payment Card Industry Data Security Standard (PCI DSS) Qualified Security Assessor (QSA) audit, for example. This audit's scope encompasses twelve areas that include both technical and non technical controls[15] and the QSA audit digs deep into each of these requirements. However, the audit is confined to the Cardholder Data Environment (CDE). If the entire corporate infrastructure is not part of the CDE, those areas outside the

---

[15] https://www.pcisecuritystandards.org

scope are not audited and therefore information security risks may go undetected.

Perhaps the most dangerous aspect of focusing on compliance is the reduction or lack thereof of a meaningful security culture. Personnel may make security-related decisions based on rules and not an understanding of the threat environment. For example, a compliant employee will recognize a phishing email and delete it, whereas a security-aware employee will forward the email to information security for further analysis, recognizing that their phish email may be a component of a broader campaign against the organization and the email could provide valuable intel for those tasked with protecting the company.

Bottom Line: Organizations that wish to improve their security posture should focus on a holistic information security program based on a recognized framework, not solely on passing audits. Compliance will follow a properly implemented and managed information security program.

SMB Considerations: When facing an audit for the first time (such as PCI DSS QSA), the tendency may be to direct all efforts to achieving only compliance. This is an incorrect approach because security is not solely about compliance. SMBs should focus on building a holistic information security program, and compliance will follow as a result.

Governance

An effective information security program begins at the top of the organization. Information security is a business concern, not limited to technology or compliance. As such, it must be addressed at the C-suite and Board of Directors levels.

Reporting Structure

A constant debate in information security is who should the (v)CISO[16] report to? In 2019, a survey found that nearly 45 percent of CISOs reported to the CIO.[17] This should not be a surprise since information security is often thought of as an information technology issue. Most of the first information security controls were information technology-related, the emergence of the corporate firewall in the 1990s a prime example.

This reporting relationship can cause significant issues, however. First, as previously stated, information security is not exclusively an information technology issue, but certainly information technology security is a subset of information security. By placing the information security role within information technology, the incorrect

---

[16] I use the notation (v)CISO in this document to call out actions that may be performed or related to both a CISO and a virtual CISO.

[17] https://www.csoonline.com/article/3278020/does-it-matter-who-the-ciso-reports-to.html

perception of information security as an information technology role is strengthened.

Second, such an organizational flow creates the possibility for conflict of interest. The CIO is responsible for keeping bits and bytes flowing. Sure, part of that is maintaining the confidentiality, availability, and integrity of those bits and bytes, but the primary concern is ensuring that the organization and its outside relationships have access to information and resources to conduct business. After all, most companies are not in the business of providing information technology or security services; those functions represent a cost of doing (and continuing to do) business. When budgets are tight, the temptation to go light on security and instead focus on streamlined operations is real.

Information security is, at its core, risk management, going well beyond information technology security. An excellent way to explain and visualize this is via the Three Lines of Defense (3LoD) model.[18] With this as a guide, information security functions may be categorized as follows:

- First Line—Operational Management (for example, information technology security controls such as

---

[18] https://www.theiia.org/en/content/position-papers/2020/the-iias-three-lines-model-an-update-of-the-three-lines-of-defense/

firewall installation, configuration, and maintenance via proper change management)

- Second Line—Risk Management, to include periodically performing a broad risk assessment to identify vulnerabilities in systems and processes that cybercriminals may be able to exploit (for example, examining firewall rules and configuration for business applicability and currency)
- Third Line—Audit (for example, providing assurances that proper firewall management occurs and aligns with corporate needs by examining evidence of such)

This model provides for proper segregation of duties, as the risk management function operates independently from the operational management function. The chances of implementation of unauthorized firewall rules are significantly reduced as the risk is properly managed through change management, reviews, and other controls.

Note that unauthorized firewall rules are not limited to intentional acts. The majority of unauthorized or risky rule changes stem from lack of knowledge of the risks involved. An example is a firewall administrator opening up SSH globally in and out instead of allowing only necessary source and destination IP addresses. The firewall administrator has successfully performed their information technology mission of keeping bits and bytes flowing, without addressing the significant business risks of allowing

TCP port 22 communications to every machine on the internet.

Another reason for maintaining 3LoD discipline is to act as an interpreter between operational management (not limited to information technology) and information security auditors. A common example is when an auditor determines a control or criterion of a certain regulation, standard, or framework is not met.

An essential part of the third line of defense is the independence of the auditor. Whoever performs the audit function should not be beholden to information technology or information security, reducing the possibility of the auditor being directed to downplay certain risks or "look the other way."

Remember, the goal here is not to check audit boxes but to reduce risk. The risk management function can effectively explain how compensating controls effectively mitigates the risk or requirement of concern without requiring the primary control.

That's not to say that information security can never report to the CIO. The (v)CISO's job is to help provide solutions to business problems that reduce risk to an acceptable (if not the lowest) level. This includes reporting. If other compensating controls such as oversight are in place, the conflict-of-interest risk of information security reporting to information technology can be effectively minimized. This is a cultural decision and should be made by the top executives, not by information technology.

Committee and Board

As a business function, information security cannot operate in a vacuum. Input from senior leaders and business executives is necessary to ensure information security aligns with the business needs.

A common practice to achieve this communication is to create an information security advisory group or steering committee. This committee is tasked with communicating information security threats and opportunities. This is not the audience to present details about vulnerability scans to; they don't need to know about the minutia of port scanning, for example. Rather, this is where the (v)CISO puts their interpretation skills to use by explaining in business risk terms what, for example, an increase in the number of vulnerabilities from the last scan means for the business and what the options are for remediation.

The audit function, described in the preceding section, should report directly to the board or to the information security committee. For companies lacking a board of directors, the audit function should report directly to company owners.

Committee meetings are typically quarterly but could occur at a different cadence so long as objectives are reached. Those objectives should be spelled out in a charter approved by the committee and, if board appointed, by the board of directors. Minutes should be taken for board briefings and as evidence for auditors of topics discussed at committee meetings.

For some organizations, these committee meetings are not enough to effectively brief the board of directors. Some boards are highly interested in information security issues, while others see it as nothing more than a compliance nuisance (not the correct approach). For the former, the (v)CISO may have the opportunity to brief and interact with the board.

For those (v)CISOs fortunate enough to get in front of the board, it is usually once a year and for a limited time—ten minutes, to include questions and answers, is common. The (v)CISO's message must be concise and on target. Slides should augment the presentation and not be "busy," and the (v)CISO should never read from the slides. Discuss relevant metrics, threats, needs, and risks, and how they tie into the business' strategic plan.

Business Unit Meetings

While executive and board participation is necessary for a well-functioning, holistic information security risk management program, communication does not stop there. The heart and soul of the business (and therefore the greatest information security threats and opportunities) are the units that create the business' deliverables and the back-office units that support them.

It is common for executives to have a one-hundred-day strategy or plan when they begin working with an organization, and the (v)CISO is no exception. Of primary importance to that plan is understanding the business as best as possible. This includes creating relationships with business units across the enterprise. The (v)CISO should plan to meet with each business unit head during this period. The meetings should tackle the business processes from a high level, what confidential information is involved (including its location), and any existing information security concerns. The (v)CISO should not come into the meetings with preconceived answers; here an open mind is required, as incorrect baseline assumptions can breed other more critical incorrect assumptions.

A second reason for the establishment of these relationships is to build trust. The business unit should not treat the (v)CISO as a technical resource or an adversary. The business unit should instead recognize the (v)CISO's goal is to help the business unit reduce risk, because the business unit, not the (v)CISO, owns the risk.[19]

Additionally, the business unit, and all employees, can serve as eyes to help identify information security risks. The (v)CISO cannot assist in risk management with unknown

---

[19] See the Risk and Risk Assessments section.

risks. Risk management involves everyone in the organization.

Staff

Staff refers to all workers in a company, to include employees and contractors. While employees and contractors have different legal and tax implications to the organization,[20] for the purposes of information security risk management they could perform similar operations and therefore expose the organization to similar risks.

Staff are often the weakest link in the information security risk management chain. For example, phishing emails[21] are prevalent because they work. While network defenses such as spam filters catch most of these attempts, they will never be able to stop them all. Network defenses are primarily reactive because they rely on known data and examples, and therefore will lag the introduction of a threat.[22]

Staff can also become the strongest link in that chain, and that journey begins with awareness of threats. Staff should

---

[20] https://www.irs.gov/businesses/small-businesses-self-employed/independent-contractor-self-employed-or-employee

[21] https://www.merriam-webster.com/dictionary/phishing

[22] More network defense products are incorporating Artificial Intelligence to make predictive decisions about possible threats—see https://www.forbes.com/sites/louiscolumbus/2019/07/14/why-ai-is-the-future-of-cybersecurity/

understand and acknowledge relevant corporate information security policies such as acceptable use.[23] They should understand their role in the business as well as the classification of information they deal with.[24] Finally, they should receive information security training commensurate to their position and the information security risk that position presents to the organization.[25]

Employee handbooks and contractor agreements must address confidentiality of information. Staff should sign and acknowledge these and/or a non disclosure agreement. This sets expectations about what staff may and may not do with your company's information.

Bottom Line: Information security is a risk management function and should be treated as such, following the 3LoD model. That includes ensuring reporting structures allow for proper risk management. Everyone has a role in information security; it is not just the responsibility of the (v)CISO. Additionally, everyone must understand their role(s) in protecting the organization.

SMB Considerations: SMBs often must cross lines of responsibility where segregation of duties is not practical

---

[23] See the Policies and Procedures section.

[24] See the Information Classification section.

[25] See the Training section.

given the size of the organization and resources available. Staff often wear many hats, making roles less distinct, especially for technical workers. One benefit of a vCISO for SMBs is the introduction of that separation (independence) in critical information security functions.

Generative Artificial Intelligence (AI)

In late 2022, ChatGPT launched and became a very popular application. In short order, many were using the generative artificial intelligence (generative AI) application for many purposes, from finding and creating recipes to asking political questions to writing policies, and more.

By the spring of 2023, several security-related issues with ChatGPT and other, similar applications emerged, namely that people were uploading confidential information as part of their analysis requests. For example, Samsung employees interacting with ChatGPT reportedly leaked sensitive corporate data. The information employees shared with the application supposedly included the source code of software responsible for measuring semiconductor equipment. The impetus for the data exposure? A Samsung worker allegedly discovered an error in the code and queried ChatGPT for a solution, providing proprietary information in the request.[26]

One may argue that the use of public generative AI applications is likely covered under existing policies such as acceptable use or information classification and handling. There is validity in this position as the more organizations write targeted policies, the more likely gaps will form in the

---

[26] https://cybernews.com/news/chatgpt-samsung-data-leak/

policy structure. For that reason, policies are constructed at a high level, and details left to procedures and standards.

But there is precedence to think otherwise. Most organizations have (or should have) a standalone social media policy. Social media use is a blend of information security, marketing, and communication concerns. Because of such, assigning social media content to one of those relevant policies may not agree with other areas' responsibilities. In other words, social media information security usage concerns could fall under the acceptable use policy, but other aspects (e.g., when, how, and what to post to social media) shouldn't be applicable to all users since only a select few should have the access and authority to post to company social media assets.

By the end of 2023, it was clear that many businesses, including SMBs, were either leveraging generative AI (knowingly or unknowingly) or had plans to do so. This was plainly evident in our business continuity tabletop exercises performed during the second half of 2023.[27] Each year, my firm scripts a different scenario, usually in the spring, that we feel is most relevant to the time. In 2023, the test scenario involved generative AI.

We noticed a few interesting trends from our test results:

---

[27] See the Business Continuity section.

- All organizations were using ChatGPT, either officially or unofficially, or both.
- Most did not know the scope of how ChatGPT was used.
- Roughly half had strategic plans to incorporate generative AI into operations.
- Approximately 10 percent had a specific generative AI policy in place.
- Most did not understand the possible security implications of generative AI use.

We recognized that, unlike the introduction of other services, the speed of adaptation and the broad, seemingly endless application of generative AI posed a significant risk to SMBs:

- Underestimating the potential threat of generative AI misuse.
- Unintentional sharing of confidential information.
- Acting on incorrect information presented by the chatbot.
- Not having a policy to direct the use of generative AI in the organization.
- Not performing a generative AI risk assessment to understand the risks.
- Not having a governance structure to direct AI strategy, risk identification and mitigation, and policy.

While there are other risks generative AI poses to small businesses,[28] addressing the ones above will minimize the risk to below the SMB's risk tolerance level—if done right.

That means choosing a framework or standard to ensure that all potential threats are holistically addressed. Just as information security programs require a framework to build on,[29] so does generative AI. Recognizing this, in 2024 my firm introduced Virtual Chief Artificial Intelligence Officer, or vCAIO services,[30] following the ISO 42001-2023 framework.[31]

Bottom Line: Generative AI is a powerful tool with great potential impact as well as risk. Therefore, it should be managed in the same manner as all else in information security, beginning with a foundational framework to build the program on. In this manner, the risks can be identified, understood, mitigated, and tracked.

SMB Considerations: SMBs may have more significant risk from generative AI use because they are less likely to have

---

[28] https://www.helpnetsecurity.com/2023/11/20/genai-usage-pressure-security/

[29] See the Frameworks section.

[30] https://vcaioservices.com

[31] https://www.ansi.org/standards-news/all-news/2023/12/12-27-23-using-ai-responsibly-us-leads-efforts-to-develop-iso-iec-42001

a holistic, structured, risk-based information security program in place. Managing generative AI risk is like managing information security risk. Those companies that put such processes in place will find adopting a generative AI framework less intrusive and cumbersome.

Information Classification and Handling

"You can't protect what you don't know about" is a common phrase in information security.[32] Additionally, you can't adequately protect what you do know about if you do not have an information classification scheme. Not all information presents the same risk—the exposure of corporate salaries produces more issues than the exposure of a public website (which, by definition, is designed for exposure).

Some use the terms "information" and "data" interchangeably, yet this is incorrect. Data is simple; data are discreet elements that, taken alone, may not reveal any confidential information. An example of data is a nine-digit number. Another example is the name of a city. Information is data with context. If the nine-digit number is paired with a name or a grade,[33] then the added context provides

---

[32] An informal search of the phrase using Google resulted in nearly 20,000 results, though it's not clear what percentage of those are information security specific.

[33] It's difficult to believe now, but more than thirty years ago it was common for university professors to post grades and Social Security Numbers on sheets of paper outside their office door. Some would even include names, presumably from printing off a roster sheet and manually writing in grades. How times have changed.

usefulness beyond the individual data elements.[34] That is why it is proper to refer to information security and not data security in most cases.[35]

Classification

An information classification scheme places information into risk buckets, which in turn aid in risk remediation by applying protective controls commensurate to the risk of information exposure or unavailability.

Information classification schemes vary, usually involving three or four classifications. For example:

- Public—All information assets that do not fit into any of the next categories can be considered public. This includes all information designed for public consumption, such as:
  - Product brochures widely distributed
  - Information available in the public domain
  - Company web sites

---

[34] There are many discussions on the differences between data and information; one is at https://byjus.com/biology/difference-between-data-and-information/.

[35] Most, but not all. Encryption in transit, for example, encrypts data, not information, since transmission is in discrete small packets, so it is proper to state "data encryption". For more information on packets and how the internet and networks operate and are designed, see https://www.amazon.com/Computer-Networks-Internets-Douglas-Comer/dp/0133587932/, a must-have reference for network professionals for decades.

- Newsletters for external transmission
- Private—Information intended for selective internal company use only. Its disclosure could adversely affect the company or its staff. Examples include:
  - Policies
  - Call lists
  - Location information
- Confidential—Information that is to be accessed on a "need to know" basis. Its unauthorized disclosure (internally or externally) could seriously and negatively impact the organization. Examples include:
  - Salaries and other personnel data
  - Personally identifiable information (PII)
  - Protected health information (PHI)
  - Cardholder data (CD)
- Sensitive[36]—Information that if disclosed could have catastrophic effects on the organization, perhaps to the point of business end, such as:
  - Confidential customer business data including contracts
  - Passwords to critical systems
  - Accounting data and internal financial reports
  - Company business plans

[36] All confidential and sensitive information collection, processing, and storage, should be for an approved and documented business reason.

The specifics of an organization's information classification may vary and are dependent on the organization's needs, information flows, and risk tolerance. The goal is to correctly apply controls commensurate to the information's risk potential.

Inventory

It may seem obvious that it's not possible to classify information without knowing what information is involved, but an information inventory goes beyond listing types of information. An effective inventory will also note all locations where the information resides, particularly for the higher-risk information.

A data flow diagram[37] is a representation of all locations where and how information is stored, processed, and transmitted. It is a useful tool for determining the presence and effectiveness of controls to protect the information. Data flow diagrams should include all locations and media where the information[38] of interest resides, to include even

---

[37] Data flow diagrams are also required by some regulations and standards (e.g., GDPR—see https://www.itgovernance.co.uk/gdpr-data-mapping and the FFIEC IT Examination Handbook—see https://ithandbook.ffiec.gov/)

[38] Note while I discuss the data flow diagrams for tracking information flow, devices and controls often act on the discrete information elements such as packets and blocks. Therefore, it may be misleading in a sense but not incorrect to use the term data for the flow diagrams.

exported Excel spreadsheet reports and paper copies, for example.

Handling

Controls exist not only in the processing, storage, and transmission of information but documenting how it is handled as well. For example, confidential information is handled in ways that maintain confidentiality, including encryption if on a thumb drive, covers if papers are left on a desk, and physical security measures if the media is transported.

Retention

Some organizations fall into the habit of keeping all information indefinitely. However, most information has retention requirements that must be met but do not have to be exceeded. Keeping information beyond what is required can introduce information security risk. For example, if your state requires medical records to be kept for seven years[39] and you have records that are older, you are taking on the risk of maintaining the information without any reward. Additionally, those records are also subject to discovery in legal action. Unless there is a justifiable, documented, and approved business reason for

---

[39] HIPAA has no PHI data retention requirements, leaving those requirements to the individual state laws and regulations. See https://www.hhs.gov/hipaa/for-professionals/faq/580/does-hipaa-require-covered-entities-to-keep-medical-records-for-any-period/index.html.

keeping records past the regulatory requirement, these records should be removed.

Owner

An information owner or steward is responsible for the use and security of the information. The business owns all information so a more descriptive title may be information steward. Regardless, usually this role is the business unit head that requires the information for the conducting of business (for example, requiring the date of birth of the customer to open a bank account for the customer). They are ultimately responsible for the security of the information—not the (v)CISO—and need to understand that responsibility.

Bottom Line: To properly address information security risk, the information must be known, classified, located, and protected according to risk.

SMB Considerations: SMBs, particularly those who are smaller and/or startups, often have complete institutional knowledge of all information inventories, locations, and controls. However, as a business grows, gaps in these can form without warning or knowledge. It is best to have a comprehensive information classification and handling process in place when small and adjust as the business grows and changes.

Policies and Procedures

A comprehensive policy suite is necessary to provide guidance and direction of the information security program. Without policies, information security governance is difficult.

Policies, procedures, and standards have different meanings:

- Policy—Defines *what* must be done, not *how* to do it.
- Procedure—Defines step by step *how* to meet the associated policy(ies) goal(s).
- Standard—Defines the specific elements allowed in *support* of the procedure(s).

For example, a network security policy may state that the network must limit access to internal systems according to industry best practices. A procedure to support that may be that firewall rules are proposed by the information technology security team, reviewed by the (v)CISO, approved through change management, and implemented during the organization's published maintenance window. The standard might be how the firewall rule is formatted and documented, the type of change request, and perhaps the approved equipment (e.g., small business firewall if setting up a branch office). Policies should only contain policy statements. All documents may include roles as well. A RACI[40] matrix may promote role clarity.

A word of caution regarding generative AI and policies. "In generative AI, algorithms are designed to learn from training data that includes examples of the desired output. By analyzing the patterns and structures within the training data, generative AI models can produce new content that shares characteristics with the original input data. In doing so, generative AI has the capacity to generate content that appears authentic and human-like."[41] Because of this, there is a great temptation to create information security policies using generative AI. However, as in the case of policy templates, if pursuing this route, organizations should look at the generative AI output as the beginning, not the end. Build on that foundation by customizing the generated policy, ensuring alignment with business needs and operations.

The exact policies needed vary from organization to organization depending on its characteristics, operations, and goals. For example, if remote work is not permitted, a remote access policy is not necessary. Aligned frameworks

---

[40] RACI stands for Responsible, Accountable, Consulted, Informed. For more information see https://racichart.org/.

[41] https://www.techrepublic.com/article/what-is-generative-ai/

(e.g., CMMC[42]) also inform policy requirements. Yet there are some core policies that every organization needs.

Information Security Policy

The information security policy provides oversight of the program. It defines the roles discussed earlier[43] as well as others depending on the organization. As such, the information security policy should define all governance and roles.

Some organizations prefer to have one policy that contains all elements of the program. That can become difficult to manage, however. A more common approach is to create several policies, each written for its intended audience. For example, someone in the accounting department likely doesn't need to know about encryption algorithms or EC2 instances.

Acceptable Use Policy

The acceptable use policy defines what actions and uses of corporate information technology and information assets are permitted. While most staff intuitively understand about protecting confidential information and not misusing assets, without concrete rules there are grey areas that may create ambiguity and risks. For example, staff should never

---

[42] https://dodcio.defense.gov/CMMC/Model/

[43] See the Governance section.

use their corporate email account for anything but corporate business, nor should they use their corporate password for other systems.

The acceptable use policy needs to be reviewed and acknowledged by all staff on an annual basis. That's not to imply that staff do not need to review and acknowledge other policies, but this is one policy intended for all staff. Policy acknowledgement can be through an HR system, a training system,[44] or some other method.

Password Policy

While password attributes such as length and complexity can be addressed as a standard, some auditors prefer that the specifics are included in a password policy. Items include length, complexity, history, and lockout requirements.

Information Classification and Handling Policy

The need for, and elements of, an information classification policy were presented earlier.[45]

---

[44] For example, KnowBe4 allows for corporate policy review and acknowledgement as part of the annual training cycle—see https://www.knowbe4.com/.

[45] See the Information Classification and Handling section.

Incident Management Policy

How incidents are handled should be determined and tested before incidents occur. This includes roles and responsibilities (possibly a RACI matrix), notification requirements, and escalation procedures. This policy differs from the business continuity and disaster recovery policy (see below) in that it's more focused on information security incidents that may or may not cause a business disruption or loss of confidential information. It should include a method for ranking and tracking incidents.

Business Continuity and Disaster Recovery Policy

A high-level business continuity and disaster recovery policy (BC/DR) directs the creation and testing of the processes necessary to continue business operations in the event of a disaster or other interruption. The business continuity procedure and standards (sometimes collectively referred to as the business continuity plan, or BCP) outlines the steps for restoration of services and the business impact analysis (BIA) of service disruption. The BCP could be quite lengthy, incorporating BCPs and BIAs for each individual business unit.

The disaster recovery plan, or DRP, outlines the steps needed to restore infrastructure and is driven by recovery time objectives and recovery point objectives (RTO, RPO) provided by the BCP and BIAs. Disaster recovery is an information technology function while business continuity is an operations or risk management one. Information technology should not make assumptions on which service restorations to prioritize; that follows from the BCP. The BCP should be tested at least annually.[46]

Encryption Policy

Like the password policy, auditors often like to see a separate encryption policy to show that information is protected at rest and in transit. The policy can be short and written at a high level, noting for example industry-standard encryption such as AES-256 and TLS1.3 is used. The procedure and standards are where details of encryption algorithms and key and certificate management reside.

User Access Policy

A significant information security risk is in the relaxing of the principle of least privilege, where staff and applications are provided only the level of access necessary to perform their functions. This can happen with incorrect assignment of permissions to staff or not removing permissions when changing roles or terminating.

There should be a formal process for approving user access. As noted previously,[47] information owners or stewards are responsible for the security of the information. Therefore,

---

[46] See the Business Continuity, Disaster Recovery, and Incident Response section.

[47] See the Information Classification and Handling section.

they must have a role in the approval and oversight of access to systems containing that information.

Two other roles play a prominent position in user access management, information system owner and user access administrator. The information system owner is the system administrator of the system to which access is requested and can provide technical details on how to accomplish the access to conform with the principle of least privilege. The user access administrator is the one who grants the specific access. While these roles may be accomplished by the same person, to maintain separation of duties, the information owner or steward role should never provide access, only authorize.

The policy should also contain a provision for review of access. Often the (v)CISO is responsible for periodic user access reviews to provide independence, though the information system owner or user access administrator should also perform the function as part of regular administration duties.

Generative Artificial Intelligence Policy

Generative AI has become a powerful business tool. From policy base creation to coding examples, the uses for this new and exciting technology seem endless. However, use comes with risks, and the need for direction.

In 2023, most of our clients either created a separate generative AI policy or incorporated elements into other policies, such as the information security and acceptable use policies. This included use of external chatbots (e.g., ChatGPT), internal LLMs for process improvements, and vendor management as part of vendor due diligence. How your organization governs generative AI is based on what you plan to use it for, but implementing guardrails on its use is necessary to reduce risks to information security through its use.[48]

Policy Approval

Policies should be reviewed and approved by the information security governance committee, the enterprise risk management committee, or similar governing body. The policy owner(s) and enactor(s) should not approve policies.

Policy Register

A policy register is a method for tracking policies and their purpose, owner, location, and last review date. It can be as simple as a spreadsheet or part of a Governance, Risk, and Compliance (GRC) system.[49] Without a register, ensuring policies are reviewed at least annually can be difficult and inefficient.

---

[48] See the Generative AI section.

[49] See the Governance, Risk, and Compliance section.

Policy Template

Policies should follow a standardized format. They should include the owner, last approval date, approver, policy number, and any referenced documents.

There are sites and services that offer templates or will create policies for your organization. Caution should be exercised if using these services. Policies that are, for the most part, copied from another source and labeled with the organization's name without incorporating organizational processes, needs, and culture may not have significant use beyond meeting a compliance requirement.

Additionally, a common tactic for due diligence reviewers is to search random passages from policies to determine if policies were copied. That's not to say that templates are not of value. A search could return common language by virtue of policies designed to align with specific frameworks that have been adopted to the business. If the policies are defensible, that is, they can be shown to be representative of processes in place, then shared template language is acceptable.

Bottom Line: Information governance and management is possible only with a set of high-level policies providing definition and guidance to the organization. These policies should be reviewed and approved by an executive-appointed committee.

SMB Considerations: SMBs often create policies in a disjointed manner whenever one is required (e.g., to convert a prospect into a client by meeting the prospect's information security due diligence needs). A short yet comprehensive policy suite to include policy registration and formalized review and approval processes will pay dividends as the business grows.

Frameworks

An information security framework provides controls to mitigate information security risks across multiple domains (areas). Security executives usually align their organization's information security program to one or more frameworks. Doing so provides a roadmap for implementing and measuring the program, provides a common language that other security professionals understand, allows for tracking of changes to aspects of the program, and demonstrates the program's maturity to existing and potential partners and clients.

The most common frameworks in use by SMBs are:

- NIST Cybersecurity Framework
- CSA Cloud Controls Matrix
- CIS Critical Security Controls

NIST Cybersecurity Framework

The National Institute of Standards and Technology Cybersecurity Framework, more commonly known as NIST CSF, was published in 2014, with a major revision in 2024, and consists of standards, guidelines, and best practices to manage cybersecurity-related risk.[50] The CSF version 2.0

[50] https://www.nist.gov/cyberframework

consists of six functions with twenty-two control categories:

- Govern
- Identify
- Protect
- Detect
- Respond
- Recover

Cloud Security Alliance CCM

The Cloud Security Alliance Cloud Controls Matrix, more commonly known as the CSA CCM, focuses on cloud computing environments. The framework is designed to augment organizations' internal security standard(s).[51]

The CCM consists of seventeen control domains that cover 197 controls. For SMBs, a light version is available that covers ninety-one controls.[52]

[51] https://cloudsecurityalliance.org/working-groups/cloud-controls-matrix/#_overview

[52] https://cloudsecurityalliance.org/research/cloud-controls-matrix

Center for Internet Security Critical Security Controls

The Center for Internet Security Critical Security Controls, more commonly known as the CIS Controls,[53]combines and consolidates CIS controls by activities, regardless of who manages devices. CIS version 8 includes eighteen controls.

Others

Other frameworks include:

- NIST 800-53A Rev 5[54]
- NIST 800-171 Rev 1[55]
- Cybersecurity Maturity Model Certification (CMMC)[56]
- ISACA COBIT 5[57]
- FFIEC Cybersecurity Assessment Tool (CAT)[58]

---

[53] https://www.cisecurity.org/controls/cis-controls-list

[54] https://csrc.nist.gov/pubs/sp/800/53/r5/upd1/final

[55] https://csrc.nist.gov/publications/detail/sp/800-171/rev-1/final

[56] https://dodcio.defense.gov/CMMC/

[57] https://cobitonline.isaca.org/getting-started

[58] https://www.ffiec.gov/pdf/cybersecurity/Cybersecurity%20Assessment%20Tool%20Slides_June_30_2015.pdf

- HITRUST CSF[59]

The Payment Card Industry Data Security Standard (PCI DSS) also provides a framework and is covered under the Regulations and Standards section.

Which Framework to Choose

Choosing a framework should not be arbitrary and should align to business needs. Frameworks are also not mutually exclusive, as they map well to each other, with some overlaps. For SMBs, often the CIS controls form an adequate base for building a security program.

The framework choice is not the end of information security decisions; rather it's the beginning. Once a basic framework is achieved, organizations often adopt other frameworks or, as in the case of some such as CMMC, strive to reach a higher level of maturity. But it is not only the adoption of other framework elements that drives continuous framework evaluation; systems, business needs, and threats are not static and change often. What may be an effective framework with associated controls today may be insufficient a year from now.

Bottom Line: The choice of framework should reflect the business needs of the organization. This is a common theme among many information security program aspects.

---

[59] https://hitrustalliance.net/

SMB Considerations: Unless there are specific regulatory requirements, the NIST CSF or CIS Controls are excellent starting points that other frameworks can build on. However, in cases where a targeted framework is required (e.g., CMMC for government business or FFIEC Cybersecurity Assessment Tool for financial institutions), start at the lowest maturity level required, whether by contract, inherent risk, or other factors, and grow from there.

Regulations and Standards

Many, if not most, SMBs are affected by at least one information security regulation or standard. A regulation is a law that must be followed whereas a standard may not have any legal repercussions for non compliance but may result in other negative effects (losing the ability to take credit card payments for PCI noncompliance, as an example).

HIPAA

The Health Information Portability and Accountability Act (HIPAA) focuses on personal health records. Hospitals and doctors' offices are obvious examples of businesses that fall under HIPAA regulations. Others may include banks (lockbox operations), counselors (psychological records), and CPAs (medical bills).

A business' primary service does not have to be directly related to health to be affected by HIPAA.

HIPAA's Privacy Rule[60] details what may and may not be done with the information collected. The HIPAA Security Rule establishes a national set of security standards for protecting certain health information that is held or

---

[60] https://www.hhs.gov/hipaa/for-professionals/security/laws-regulations/index.html

transferred in electronic form and thereby enforces elements of the Privacy Rule.

HITRUST is the information security framework for HIPAA compliance and maps to and complements other frameworks. Several firms offer audits to demonstrate HITRUST compliance.[61]

GLBA

The Gramm-Leach-Bliley Act (GLBA) requires financial institutions to explicitly describe how information is shared and to protect confidential consumer financial information A financial institution is a company that offer consumers financial products or services like loans, financial or investment advice, or insurance.[62] The FFIEC provides a framework for complying with the information security aspects of the GLBA.[63]

GDPR

The General Data Protection Regulation (GDPR) (EU 2016/6791) regulates the processing by an individual, a company, or an organization of personal data relating to

---

[61] https://hitrustalliance.net/csf-assessors/

[62] https://www.ftc.gov/tips-advice/business-center/privacy-and-security/gramm-leach-bliley-act

[63] https://ithandbook.ffiec.gov/

individuals in the European Union (EU).[64] The regulation affects organizations both within and external to the EU. Therefore, US-based companies that process EU citizen personal data should ensure that their processes comply with the GDPR. The GDPR consists of ninety-nine articles that outline privacy and security requirements.

PCI DSS

The Payment Card Industry Data Security Standard (PCI DSS)[65] specifies controls for securing Cardholder Data. PCI DSS consists of twelve requirements. The level of compliance necessary for each requirement is determined by an organization's transaction volume. PCI DSS compliance is often self-attested by SMBs, but an audit process by qualified assessors (PCI QSA) may be required for larger organizations.

Other

There are many other regulations and standards, such as individual state reporting requirements. The (v)CISO, legal, finance, and other resources can determine which are applicable.

---

[64] https://ec.europa.eu/info/law/law-topic/data-protection/reform/what-does-general-data-protection-regulation-gdpr-govern_en

[65] https://blog.pcisecuritystandards.org/at-a-glance-pci-dss-v4-0

Bottom Line: Frameworks help to ensure compliance with regulations and standards, and the information security program development should follow that path.

SMB Considerations: Sometimes compliance becomes the initial driver for starting or enhancing an information security program. In that case, care should be taken to choose the applicable framework that most closely matches the compliance requirements the organization must follow (e.g., HITRUST for HIPAA).

Maturity Assessments

A maturity or readiness assessment measures an organization's alignment with the requirements of a framework, regulation, or standard by examining the controls in place to meet these requirements. Typically, when speaking about controls, look at three categories, presented in the Basic Information Technology Control section and expanded here:

- Preventive—Controls that prevent an actual or potential vulnerability from being exploited. Patch management (technical), an acceptable use policy (managerial), and a door lock (physical) are examples.
- Detective—Controls that identify vulnerabilities. Vulnerability assessments and penetration testing (technical), phishing exercises (managerial), and surveillance cameras (physical) are examples.
- Corrective—Controls to minimize the risk of exposure (loss) once a vulnerability has been exploited. Backups (technical), a termination policy (managerial), and sump pumps (physical) are examples.

Maturity assessments are often incorrectly referred to as risk assessments. While conducting an appropriate maturity assessment can, in some cases, satisfy regulatory and standard requirements for a risk assessment, there are differences.[66]

Examples

Frameworks

- The Cybersecurity and Infrastructure Security Agency (CISA) Cyber Resilience Review (CRR) aligns to the NIST CSF. The CRR "assesses enterprise programs and practices across a range of ten domains including risk management, incident management, service continuity, and others. The assessment is designed to measure existing organizational resilience as well as provide a gap analysis for improvement based on recognized best practices."[67]
- The Educause Information Security Program Assessment Tool is a self-assessment based on the ISO 27001 framework. It is designed for higher education institutions. It includes mapping to NIST 800-53, NIST CSF, and the CIS Critical Controls.[68]

---

[66] See the Risk and Risk Assessment section.

[67] https://www.cisa.gov/resources-tools/services/cyber-resilience-review-crr

[68] https://library.educause.edu/resources/2015/11/information-security-program-assessment-tool

- The Cloud Security Alliance Cloud Controls Matrix spreadsheet is a self-assessment tool for the CCM.[69]

Regulations/Standards

- The FSSCC Automated Cybersecurity Assessment tool helps financial institutions "identify their risks, assess their cybersecurity preparedness, and help inform their risk management strategies."[70]
- The Security Risk Assessment (SRA) Tool "is designed to help healthcare providers conduct a security risk assessment as required by the HIPAA Security Rule and the Centers for Medicare and Medicaid Service (CMS) Electronic Health Record (EHR) Incentive Program."[71]
- The ISACA GDPR Assessment, consisting of 46 questions, "is a tool designed to provide a quick evaluation of an enterprise's functions, processes and procedures that indicate compliance with the European Union Global Data Protection Regulation."[72]

---

[69] https://cloudsecurityalliance.org/research/cloud-controls-matrix

[70] https://fsscc.org/

[71] https://www.healthit.gov/topic/privacy-security-and-hipaa/security-risk-assessment

[72] https://gdprassessment.isaca.org/login

- The PCI Security Standards Council's Self-Assessment Questionnaire (SAQ) is "a self-validation tool to assess security for cardholder data"[73] and contains a series of yes or no questions.[74]

Often, businesses are subjected to more than one framework and/or standard. The control requirements for each often overlap. It would not be practical to create corporate control for each requirement if in fact they are essentially the same (e.g., requiring encryption of confidential information at rest).

To address this, there are several mapping resources, where similar controls are identified across standards. One of the most useful is the Secure Control Framework[75] or SCF. The SCF is a "meta framework," or a framework of frameworks. It maps to dozens of frameworks and standards.

---

[73] https://www.pcisecuritystandards.org/pci_security/completing_self_assessment

[74] PCI DSS 4.0 became effective in 2024 and includes other responses to provide a more fine-grained view of the Cardholder Data Environment.

[75] https://securecontrolsframework.com/

Assessment Process

The maturity or readiness assessment process schedule is scope dependent. A typical assessment may consist of approximately sixty to 120 hours over three to four months. This can vary, however, based on the status of the organization's security program and the resources it is able to commit. A more mature program may take fewer hours over a shorter period. In all cases, a successful assessment requires the constant engagement of the organization. A proper assessment cannot be performed by simply providing paperwork to the assessor.

Generally, the assessment process will occur in several phases.[76]

Determine Scope

The scope of the assessment must be defined prior to beginning the engagement to properly estimate resources necessary and create an optimum schedule. Scope goes beyond the standard assessed to—it includes what aspects of the organization are to be examined. For example, PCI-DSS is only relevant to the portions of the infrastructure and personnel that handle Cardholder Data (referred to as the Cardholder Data Environment, or CDE).

---

[76] This process is valid for a variety of purposes, including procuring audit and vCISO services.

Scope may be limited to one or more of:

- Framework, regulation, or standard
- Business unit
- System(s)
- Geographic area

Find Assessor

Finding an assessor can be a challenge. This is not an audit, and requires an organization with significant risk management and security experience. Audits often focus on controls one at a time because that is the audit function—ensuring an organization is doing what they say they are doing. Risk assessors examine the threat environment and recommend the implementation of controls that the auditors then assess. In other words, a risk assessor will more likely provide several different strategies for mitigating a risk, while an auditor will measure the effectiveness of the strategy in place.

The person in the role responsible for information security in the organization should serve as the project sponsor and therefore conduct the search for an assessor. Choosing an assessor without conducting proper due diligence may result in a substandard assessor. The lead assessor should have at least five years of experience as a senior information security executive to understand the business' environment and needs beyond technical controls.

At minimum, discuss the engagement with the organization (most will provide fifteen to thirty minutes of free consulting time to both assist you in the process and learn more about the project needs). Additionally, consider reference checks from current or previous clients.

Create and Evaluate the RFI or RFP

A Request for Information (RFI) or Request for Proposal (RFP)[77] are formal processes to solicit providers for information on a service or product to solve a business issue. Once the assessor is chosen, the organization should provide the details necessary for the assessor to create a Statement of Work (SOW) all agree on and is in agreement with the RFP criteria. Usually, execution of a Mutual Non Disclosure Agreement (MNDA) prior to this information share is advised. The assessor requires access to company confidential information to properly scope the SOW.

The evaluation of the RFP should include ensuring that all requested elements were answered. Pricing should be straightforward, including additional costs if scope is exceeded and if travel is required. The experience and background of the resources should match the need (someone without healthcare experience should not be tasked to perform a HITRUST gap assessment).

---

[77] https://www.techopedia.com/definition/9167/request-for-proposal-rfp

The Engagement

Following the acceptance and execution of the MNDA, SOW, and contract (to govern the terms of the engagement), a kickoff meeting is held. This should be scheduled for the beginning of the engagement period and include Subject Matter Experts (SMEs) and other stakeholders from all relevant areas. The assessor will usually inform the organization what business units and functions should be represented.

In parallel to the kickoff meeting, the assessor will likely provide an Initial Request List (IRL).[78] This includes requests for policies, procedures, tests, audit findings, and other documentation relevant to the assessment. It is critical that the organization provide these as quickly as possible to aid the assessor in learning about the business environment. Remember, the assessor is not familiar with your organization's operations, and needs to come up to speed rapidly to properly evaluate the environment and provide the most accurate assessment. All information should be uploaded to a secure portal (usually provided by the assessor) or provided via another secure method. If the assessor suggests sending confidential information via

---

[78] Some organizations refer to this as an Information Request List. The goals are the same.

unsecured email, you may want to consider another assessor.

As part of the IRL, owners of documentation and processes are requested. After the first review of provided documentation, the assessor will schedule interviews with the relevant SMEs and other stakeholders. Questions may not be limited to these interviews; the assessor may have follow-up items. The interview typically lasts between thirty minutes and one hour but may be longer depending on the topic.

As the evidence is provided and the interviews are conducted, the assessor will begin the assessment process. The assessor's goal is to determine methods by which the agreed-on requirements are met. The assessor may leverage a dashboard or other tracking tool. If the assessor cannot determine that a requirement is met to their satisfaction, they may request a follow-up interview or email to answer additional questions. The assessor's goal is to ensure risks are addressed, not to complete a checklist.

Deliverables

Once the assessment is complete, the assessor will produce a draft report for the organization to review. This is a critical process, requiring careful examination by the organization. As the assessor has only limited exposure to the company environment, gaps may be addressed in ways that the initial assessment did not identify.

Once the edits are complete, the assessor will produce the final report, including high-level recommendations for addressing the gaps. Performing or assisting in these remediations is outside the scope of an assessment.

Finally, an exit meeting is scheduled, usually with the same SMEs and other stakeholders as the kickoff meeting, to discuss any remaining questions about the report and the organization's environment.

Bottom Line: Maturity assessments help an organization gauge where they stand in relation to a particular regulation or standard.

SMB Considerations: SMBs benefit from maturity assessments by understanding what is needed in terms of resources and commitment to reach a desired maturity level of compliance. While compliance does not equate to security, it is often required in contracts, and can be a competitive differentiator.

Vulnerability Assessments and Penetration Tests

Vulnerability assessments are automated processes to determine potential infrastructure exploitation points, such as a machine not patched for a high-risk virus. Penetration tests are human-driven authorized attempts to exploit vulnerabilities. Internal and external vulnerability assessments should occur at least quarterly, and penetration (pen) tests annually.

Consecutive vulnerability assessments should not reveal identical vulnerability points if the vulnerability mitigation program is effective. There may be valid business or technical reasons for not remediating the vulnerabilities. In these cases, exceptions must be authorized and documented (often by a quarterly steering committee).

Penetration tests are another measurement of a vulnerability management program. Penetration tests take vulnerability assessments a step further with human interaction to attempt to exploit vulnerabilities. When engaging a firm to perform penetration tests, the scope must be understood by all parties. Often the penetration tester is authorized to show they could exfiltrate data without doing so, thereby reducing the risk of an unintended breach.

A Web Application Scan (WAS) attempts to discover OWASP Top Ten[79] and other vulnerabilities in web servers. These vulnerabilities are commonly exploited to retrieve confidential information or cause a Denial-of-Service condition. Other assessments such as code scanning can detect programming errors, either in production or pre deployment.

Vulnerability assessments can be non credentialed or credentialed. The difference is whether the scanning software has the credentials to authenticate to a service. This is important in that attributes such as web pages often cannot be accessed unless authenticated to the server.

Credentialed access can be username and password and may use a certificate[80] or leverage MFA. Care must be taken when permitting credentialed scans, as the scanning service will have access to all resources those credentials have access to.

---

[79] https://www.owasp.org/index.php/Category:OWASP_Top_Ten_Project

[80] "A digitally signed document that serves to validate the sender's authorization and name." For more detailed information, see https://www.techtarget.com/searchsecurity/definition/public-key-certificate

Vulnerabilities discovered are scored, usually using the Common Vulnerability Scoring System (CVSS).[81] From this, vulnerabilities can be categorized based on risk. Organizations should have a documented policy to direct remediation times based on criticality, for example:

- Critical—Immediately
- High—Within thirty days
- Medium—Within ninety days
- Low—As necessary

Any vulnerabilities not remediated should have a documented exception. Risk acceptance is a valid risk management strategy, so long as the decision-makers have enough information to make a risk-informed decision.[82]

Penetration tests can include extracting confidential information or proving that extraction can be accomplished, depending on the risk to and the risk tolerance of the organization. Proper scoping is critical for all parties to understand the potential ramifications of penetration testing. Often it is preferable to keep some target systems out of scope; these should be documented and agreed on prior to the test beginning.

---

[81] https://nvd.nist.gov/vuln-metrics/cvss

[82] See the Risk and Risk Assessment section.

Vet your scanning provider carefully. Too often vulnerability assessments are presented as penetration tests. An example of one especially disturbing case was a so-called external penetration test that was nothing more than a Greenbone[83] scan improperly set so that the scanner could not accurately scan the targets.[84] Yet the test provider presented the results as if they conducted an extensive review and found that there were no exploitable vulnerabilities. Subsequent scans by a reputable service revealed several unknown (and therefore unaddressed) high-risk vulnerabilities.

Bottom Line: Vulnerability assessments and penetration tests, performed by a skilled and reputable organization, provide an organization information to proactively address vulnerabilities before cybercriminals exploit them.

SMB Considerations: Even purely Software as a Service (SaaS)[85] organizations should consider some form of assessments and testing, even if limited to WAS or code

---

[83] https://www.greenbone.net/en/

[84] Review of the "penetration test" report pointed to the likelihood that the scanner was set to some discovery method such as ICMP Echo that did not return any live targets because of firewall or ACL policy. A good practice is to set the scanner to always assume the targets are live.

[85] A software distribution model in which a cloud provider hosts applications and makes them available to end users over the internet. https://www.techtarget.com/searchcloudcomputing/definition/Software-as-a-Service

scans. They must carefully vet the scanning provider, though, to ensure that the provider's service is ethical.[86] Ask for all documentation, including how scans were configured and specific actions performed by the penetration tester.

---

[86] See the Ethics section.

Threat Intelligence

The term "threat intelligence" may conjure up images of government agents in suits and aviator sunglasses, or cybercriminals[87] wearing hoodies,[88] whose intelligence is useful to only fringe elements. However, as information security is a business issue, understanding the threat environment is necessary for making risk-informed decisions. There are many threat intelligence sources, including media, sector-specific Information Sharing and Analysis Centers (ISACs), and communities including professional organizations.

Media

Media is often a good source for cybersecurity threat intelligence. For example, Brian Krebs, former investigative journalist for the Washington Post, has used his skills in uncovering cybersecurity threats for many years.[89] CSO

---

[87] Many use the term "hacker" (incorrectly in my opinion) to refer to cybercriminals. The term originally referred to one who excelled at programming. See https://techterms.com/definition/hacker.

[88] The "hoodie hacker" representation feeds into this perception and obfuscates the reality that most cybercriminals are part of an organization with a goal. See https://youtu.be/ccf1zW8HeMg for a more detailed explanation.

[89] https://krebsonsecurity.com/

Online[90] is another good source for current cyber threat news. Also, podcasts such as The Virtual CISO Moment[91] provide security information and strategies targeted to SMBs on a regular basis. There are many resources, some better than others. Business leaders and other staff should be familiar with one or more sources.[92]

A caution though about relying on media, particularly mainstream media, for cybersecurity threat intelligence. There is a marketing angle included in some threat analyses that, at the very least, needs to be recognized. Just because a threat does not have its own icon or logo[93] doesn't diminish its potential for damage.

ISAC

Information Sharing and Analysis Centers (ISACs) provide venues for reporting and sharing threat intelligence to include Indicators of Compromise (IOCs). There are many

---

[90] https://www.csoonline.com/

[91] https://vcisopodcast.net/

[92] Using an RSS browser plug in is one efficient method for collecting media threat intel. Many feeds, updated regularly, are available at https://blog.feedspot.com/cyber_security_rss_feeds/.

[93] Heartbleed may be the most well-known example of this, and for good reason—it is a serious and widespread (at the time) vulnerability. See https://heartbleed.com/.

sector-specific ISACs.[94] The first, and one of (if not the most) mature, is the Financial Services ISAC (FS-ISAC).[95]

ISAC membership is generally not free, and like other member-driven organizations, the value usually far exceeds the cost. Many also hold annual conferences, which provide valuable information and allow the opportunity for in-person networking with sector-specific colleagues.

Communities and Professional Organizations

Information-security-specific communities and professional organizations provide relationship-building opportunities, especially for cybersecurity professionals. They may be the first source of information on a threat or may offer valuable peer context about known threats. Additionally, many offer certifications to establish baseline qualifications and a knowledge threshold for practitioners.

ISC2

ISC2 offers many security-specific certifications[96] including what some consider the information security leadership standard, the Certified Information Systems Security

---

[94] The National Council of ISACs member listing is at https://www.nationalisacs.org/member-isacs.

[95] https://www.fsisac.com/

[96] https://www.isc2.org/

Professional (CISSP). They also offer a community[97] for sharing knowledge and threat intel. Local chapter events offer opportunities for members to network.

ISACA

Formerly known as the Information Systems Audit and Control Association, ISACA offers several security and privacy-specific certifications.[98] ISACA is a membership organization with hundreds of local chapters around the world, conferences, and education and professional development events.

ISSA

The Information Systems Security Association (ISSA)[99] also offers local chapters for networking and learning about threats and solutions. Some chapters conduct annual security conferences such as InfoSec Nashville[100] to further networking and learning.

---

[97] https://community.isc2.org/

[98] https://www.isaca.org/

[99] https://www.issa.org/

[100] https://www.issa.org/event/infosec-nashville/

InfraGard

InfraGard is "a partnership between the Federal Bureau of Investigation (FBI) and members of the private sector for the protection of U.S. Critical Infrastructure."[101] As with ISC2, ISACA, and ISSA, local InfraGard chapters provide networking opportunities and often include meeting with local FBI and DHS cybersecurity agents. Establishing these relationships is an important component of any organization's incident response plan. InfraGard also has a comprehensive threat intelligence portal for vetted members.

LinkedIn

While LinkedIn[102] is typically considered a business social networking platform, it also is a source of threat intelligence and security awareness through groups. There are some groups dedicated to small businesses as well as some for security, to include virtual CISO information.[103]

Bottom Line: Without understanding threats, managing information security risk is not possible. There are many

---

[101] https://www.infragard.org/

[102] http://linkedin.com/

[103] The Virtual CISO Exchange is an example where virtual CISOs and those interested in the space can network—see https://www.linkedin.com/groups/12095465/.

sources of threat intelligence not directed solely to information technology security analysts.

SMB Considerations: Small companies face most of the same threats as larger organizations and should have a commensurate awareness of those that apply to their organization.

Risk and Risk Assessments

I have discussed that information security is a risk management issue and that the Three Lines of Defense (3LoD) model is useful in information security management and governance.[104] However, managing risk requires understanding what risks are and their potential to disrupt an organization.

A risk is a measure of the extent to which an entity is threatened by a potential circumstance or event, and typically a function of the adverse impacts that would arise if the circumstance or event occurs; and the likelihood of occurrence.[105] Information security risks are owned by the business, not the (v)CISO. However, it is the (v)CISO's responsibility to provide adequate guidance and input to the business for business leaders to make risk-informed decisions.

Risk assessments go beyond maturity assessments by providing finer-grained rankings for prioritizing the mitigation of risks. No organization has unlimited resources to address risks, and processing assessment results can often be overwhelming. For example, how do you go beyond the default general ranking system of the

---

[104] See the Governance section.

[105] https://csrc.nist.gov/glossary/term/risk

vulnerability assessment to determine which vulnerabilities provide the greatest information security risks specifically for your organization?

Information security at its core is risk management. There is no environment that is completely secure if information is accessible. On the other hand, information that cannot be accessed is useless. Risk management helps find the ideal balance between security and accessibility for an organization. It depends on the organization's risk tolerance, or the amount of information security risk the organization is willing to accept.

While, organizationally, information security often falls under information technology,[106] the CIO or information technology director should not decide the information security risk tolerance of the organization. This is a business decision that needs to involve senior executives, preferably through a formal process such as an Enterprise Risk Management (ERM) committee.

There are two types of information security risk assessments—qualitative and quantitative.

Qualitative

A qualitative information security risk assessment:

---

[106] See the Governance section.

- Gathers data regarding the information and technology assets of the organization, threats to those assets, vulnerabilities, existing security controls and processes, and the current security standards and requirements;
- Analyzes the probability and impact associated with the known threats and vulnerabilities to their assets; and
- Prioritizes the risks present due to threats and vulnerabilities to determine the appropriate level of training, controls, and assurance necessary for effective mitigation.

The familiar heat maps (red, amber, green) are a product of a qualitative risk assessment.

Qualitative risk assessments often involve numbers that are not measuring any tangible unit. Rather, they are representative of the assessor's opinion based on evidence and information provided; interviews; and industry, technology, and threat environment knowledge and experience. That is one reason why having two or more assessors perform a risk assessment is not advisable—they have inherent differences in experience and that may show in their subjective ranking.

Still, the qualitative method remains the most common form of information security risk assessment today and is a feature of the frameworks previously discussed. It is a satisfactory process in determining the relative rank of high information security risks. While the order of the top five security risks of an organization may differ by assessor, chances are favorable that each list will contain the same five risks.

Quantitative

Quantitative risk assessments estimate risk exposure (potential loss). This approach is commonly used in easily quantifiable industries such as insurance. Rates are based on healthcare costs, studies of behavioral factors, indicators of health issues, and so on. Information security and privacy, however, have comparatively little accurate historical data to draw from. Some, such as eRisk Hub, have attempted to quantify the cost of a data breach in terms of records exposed based on historical data, or offer data breach calculators,[107] but these can only produce high-level approximations.

To illustrate the complexity of quantitative risk assessments, if the risk of having an insufficient training program is rated high, what does that measure mean in terms of actual loss? In other words, what is an organization's loss exposure given an assumed deficiency in information security awareness? This can only be answered by looking at many variables, to possibly include but not be limited to:

---

[107] https://eriskhub.com/mini-dbcc

- Average cost of a data breach,
- Likelihood that insufficient awareness will lead to exposing a vulnerability,
- Likelihood that vulnerability will be exploited,
- What information is exfiltrated,
- How much information is exfiltrated,
- Regulations or standards violated,
- Recovery time,
- Revenue loss,
- Capital expended in remediation actions (forensics, notifications, etc.), and
- Cyber insurance coverage.

This is not a complete list but rather shows the complexity in attempting to determine the loss exposure given an identified risk.

The Factor Analysis of Information Risk (FAIR) is an international standard model for quantifying information security risk.[108] FAIR takes input based on a detailed analysis of the organizations' threat exposure, controls in place, and potential costs. Like the qualitative model, many of the inputs are subjective to an extent, there are more opportunities for describing the actual business environment. Therefore, while more complex, FAIR may calculate a more accurate cost exposure range.

---

[108] https://www.fairinstitute.org/

FAIR at the highest level emulates traditional qualitative risk assessments by determining risk from loss event frequency (likelihood) and loss magnitude (impact). However, whereas determining likelihood and impact is solely based on the assessors' experience and opinion, FAIR incorporates more granular information to determine these, specifically threat event frequency and vulnerability for loss event frequency and threat capability and resistance strength for loss magnitude. Similarly, on the other side of the equation, loss event magnitude can be broken down into primary loss and secondary risk, and secondary risk further breaks down into secondary loss event frequency and secondary loss magnitude.

Regardless of type, risk assessments provide guidance for SMBs to understand and address identified information security risks. With such information, executive management can make risk-informed decisions regarding applying resources and strategies for mitigation. Risk mitigation options include:

- Applying appropriate control(s),
- Transferring the risk to another party such as with insurance,
- Avoiding the risk by eliminating the environment that produced the risk, or
- Accepting the risk.

Risk acceptance is a valid mitigation strategy, but only after thoroughly understanding the risk and the resources needed to reduce the likelihood and/or impact of the risk, all while considering the organization's risk appetite.

Assessment Process

The risk assessment process follows the same methodology as discussed earlier in the Maturity Assessment section. A typical assessment may take fifty to ninety hours, or more for more complex environments, over at least two to four months, and includes some or all the following phases:

- Requesting and evaluating documentation and information,
- Performing maturity analysis,
- Creating and completing the information security risk register, and
- Creating and presenting the Information Security Risk Assessment Executive Summary report.

Data inputs for the risk assessment elements include but are not limited to:

- Information security events and incidents,
- Previous audits and exams,
- Regulatory changes and additions,
- Addition of new technologies and systems,
- Changes to the threat environment,
- Service provider initial and ongoing due diligence reviews, and
- Results of the previous risk mitigation actions.

Evaluation of risk impacts considers several criteria, including but not limited to:

- Strategic value to the business,
- Information classification,
- Threat environment,
- Regulatory actions,
- Maintenance of confidentiality, integrity, and availability of information, and
- Customer relationships and service.

The risk register is perhaps the most important element of this process. Additionally, it can contain, but should not be limited to, computing assets. Information assets and processes should be part of the evaluation process.

More complex organizations may use tools to create the risk register, however such complexity is often not necessary for SMBs. Many use a spreadsheet to create the risk register. The process does not need to be complicated to be effective. Avoid letting a tool manage the process rather than allowing for the process to manage tool use.

Bottom Line: Risk assessments are often the most important tool that a (v)CISO has, as they convey both the highest risks to executive management and the board of directors and the rationale behind the ranking. Qualitative risk assessments are often a representation of an SME's opinion, whereas quantitative assessments help to unveil the cost of not mitigating the risk.

SMB Considerations: Start simple with risk assessments. Begin by asking both information technology and other business units what they think are the greatest risks to information security. Their lists will not be complete but will offer guidance to determining the existing risks in your organization.

Control Catalog

Controls are actions taken by an organization to mitigate risk. They also can meet framework, regulatory, and standard requirements. For example, a regulation may dictate an entity protects information in transit, and the entity's control is forcing TLS 1.2 and above encryption on all web sessions.

Managing controls as a discrete yet relative component is an effective way to managing the information security program. Controls can map to or have relationships with risks, frameworks, and audits.[109] Using the example above, encryption in transit (the control is in place for all networked elements transporting sensitive information) ensures that requirements of different frameworks, compensating controls for risk mitigations, and audit responses can all be addressed.

As noted earlier, there are various types of controls. Many view information security controls as only technical and preventive (such as firewall rules).[110] However, from a functional perspective, they may be physical, technical, or

---

[109] See the Governance, Risk, and Compliance section.

[110] Some postulate other categories of controls. The point here is to look at both functional and temporal aspects.

managerial. Additionally, from a temporal view, they can be preventive, detective, or restorative.

Controls should have an owner and should undergo periodic maintenance or review (at least annually). For example, a simple control may be that all confidential information is encrypted in transit. The maintenance or review[111] would be to verify that all instances of confidential information transmission are encrypted according to the organization's encryption policy.[112]

Bottom Line: Tracking controls may seem like overhead, but the benefits are realized when mapping to risk assessments, maturity assessments, and audits. If all mitigating controls identified for addressing a particular risk are designed and operating properly, the residual risk is easily shown as low.

SMB Considerations: When working with the initial risk assessment, the actions to mitigate risks become the initial control catalog. Begin creating the control catalog by documenting what is in place.

---

[111] GRC systems can track and map control maintenances and relationships—see the Governance, Risk, and Control section.

[112] See the Policy and Procedures section.

Audits

The AICPA Service Organizational Control (SOC) audits for service providers are "internal control reports on the services provided by a service organization providing valuable information that users need to assess and address the risks associated with an outsourced service."[113] There are a few categories of SOC reports:

- SOC1 (formerly known as the SSAE 16 and before that the SAS 70) focuses on controls to protect financial information
- SOC2 focuses on the information security and privacy controls
- SOC3 summarizes a SOC2 without providing confidential information
- SOC for Cybersecurity focuses solely on cybersecurity risk (a new category that has not yet seen widespread adoption)

Additionally, the SOC1 and SOC2 come in two types. Type 1 (T1) assures that the controls in place are properly designed and in place for the requirements whereas Type 2 (T2) tests the effectiveness of the controls. The SOC2 T1 is therefore

---

113 https://www.aicpa.org/interestareas/frc/assuranceadvisoryservices/socforserviceorganizations.html

a snapshot in time while the SOC2 T2 covers a certain period (between three months and one year).

The SOC2 evaluates against one to five Trust Criteria areas:

- Security (also referred to as Common Criteria)
- Availability
- Confidentiality
- Privacy
- Processing Integrity

Most organizations do not request evaluations against all five because not all are applicable. For example, a hosting environment that provides rack space, power, and network connectivity may be evaluated for Security and Availability but not for the other three since it does not have access to data.

Many SMBs elect to undergo first a SOC2 T1[114] then a SOC2 T2 audit as they grow, and the information security program matures to demonstrate their security posture to customers, prospects and other interested parties. In fact, the SOC2T2 audit is a competitive advantage, as many organizations will not enter a relationship with a service provider that does not undergo an applicable annual

---

[114] Type Two differs from Type One in that it examines evidence the controls were effective during a predetermined period, such as a year. See Third-Party Service Provider Reviews for more information on the SOC audits.

SOC2T2 audit or other third-party attestation regarding the information security program's design and effectiveness.

The SOC1 and SOC2 reports generally contain sections that address the following:[115]

- Auditor statement—The auditor's description of their audit process and scope;
- Service provider description—The service provider's statement on services offered along with other company information;
- Complimentary user entity controls—The controls the partner must fulfill to ensure correct operation of the controls tested (such as access policies);
- Controls—The description, applicability, and in the case of the Type 2, the result of the testing of effectiveness (usually no exceptions or exception noted) of the controls;
- Management response—The service provider's answer to any exceptions; and
- Other items not audited—Often the service provider will supply supporting documents here such as the business continuity policy.

---

[115] See the Third-Party Service Provider Reviews section for more information on the SOC audits.

The SOC2 audit process mirrors that discussed for previous assessments.[116] Time and cost are dependent on scope and criteria evaluated against. The audit must be performed by a qualified organization.[117]

Other audits businesses may encounter are PCI DSS QSA (if handling credit and debit card data), ISO 27001 (more common outside the United States), the FFIEC examination (for financial institutions), and HITRUST (for healthcare organizations). To prepare for an audit regardless of focus, the organization should have a clear understanding of where they stand prior to engaging an audit firm. This should involve a relevant gap assessment, either internal or external.

While auditors are tasked with pointing out deficiencies, good auditors are a resource for the business. If deficiencies are not extreme, the auditor may provide time for remediation during the audit. Regardless, findings are not necessarily looked upon as a solid negative. So long as the management response to the finding is appropriate, those evaluating the audit will likely be satisfied.[118]

---

[116] See the Maturity Assessments section.

[117] For example, SOC2—see https://www.aicpa.org/interestareas/frc/assuranceadvisoryservices/aicpasoc2report.html.

[118] See the Third-Party Service Provider Reviews section.

Bottom Line: Audits are necessary for providing evidence to outside entities that your organization adequately addresses information security risks. This independent attestation should remove the requirement for customers to perform their own audits. Additionally, third-party audits are often accepted in lieu of responding to clients' questionnaires.

SMB Considerations: Don't think your organization is too small to consider an independent audit. The smallest organization my firm has worked with to reach SOC2 T1 then T2 compliance had fewer than thirty staff. Even for small organizations, a SOC2 or another independent audit can be a significant competitive differentiator.

Third-Party Service Provider Reviews

As part of a complete information security program, third-party risk must be assessed and treated, and is referred to as Third Party Risk Management (TPRM). Migrating information to a cloud platform does not absolve the organization of the responsibility for protecting information. Therefore, Third-Party Service Provider (TPSP) information security reviews should be performed as part of initial and ongoing due diligence. Depending on the organization, these reviews may also be a regulatory requirement.

TPSP reviews begin with a vendor risk assessment to determine the critical vendors and therefore where to prioritize resources. Any service provider that stores and/or processes organizational confidential information is a critical vendor. Type, quantity, and accessibility of data further define the criticality.

A typical TPSP review may involve examining some or all the following:

- Audits
- Questionnaire responses
- Policies and procedures
- Vulnerability assessments and penetration tests
- Certifications

TPSP Audits

Audits present an independent view of the organization's infrastructure. In many cases, they are the only source of visibility into the TPSP's security posture.

Audits that show repeated findings for the same exception represent a cause for concern to the assessor. An information security program that does not proactively address gaps is more susceptible to information security risk exposure.

The two audit types SMBs will most commonly encounter for TPRM due diligence are SOC2 and ISO 27001.

SOC2

SOC2[119] reports can be qualified or unqualified. A qualified report means the auditor's opinion is "qualified" by one or more material exceptions, whereas an unqualified report is one that is considered clean (without material exceptions). The terminology can be confusing, as unqualified refers to the report's details, not the service provider.

Many who evaluate SOC2 reports ignore the Complimentary User Entity Controls (CUECs); this can be a critical error. Without verifying the CUECs are in place and function well in the organization, there cannot be assurance that all the TPSP's controls will be effective. Information

---

[119] See the Audit section for more information on SOC reports.

security risk management truly is a collaborative effort. Care should be taken to evaluating the CUECs against the organization's controls (see Control Catalog).

ISO 27001

ISO 27001 is an international information security standard. Originally released by the International Organization for Standardization (ISO) as ISO 17799, in 2005 it spawned a new major category of ISO, the 27000 series. To date there are more than three-dozen categories in the ISO 27000 series. ISO 27001 was last updated in 2023 (ISO 27001:2022).

ISO 27001 describes the requirements for the Information Security Management System (ISMS), including the control areas—four groups covering ninety-three controls:

- Organizational
- People
- Physical
- Technological

ISO 27002 provides more guidance and information for implementing the controls in the groups listed above as necessary. Organizations can become ISO 27001 certified to demonstrate compliance with the framework. This process often begins with an ISO 27002 Control Readiness Assessment.[120]

Questionnaire Responses

Many organizations create security questionnaires that may have anywhere from a dozen to several hundred questions about a firm's information security environment. The majority of these are company-dependent, though some use a standardized format such as the Shared Assessments Standardized Information Gathering (SIG) and SIG Lite.[121]

Volume and detail of questionnaires can result in slower response times. Two weeks is a reasonable turnaround expectation, though that may be impacted by several factors such as SME availability and number of requests. Both requestor and responder need to work together on the process to ensure a timely yet reasonable response.

Policies and Procedures

Policies and procedures demonstrate the structure of an organization's information security program, usually including the framework(s) it aligns to. Ideally the policies and procedures should cover all the applicable categories in

---

[120] See the Maturity Assessments section.

[121] https://sharedassessments.org/sig/

the aligned framework(s). Typically, this will involve several policies, including but not limited to the following topics:[122]

- Information security governance and roles;
- Acceptable use of assets;
- Password creation, structure, and management;
- Information classification and handling;
- Incident management;
- Encryption use and requirements;
- User access management; and
- Business continuity and disaster recovery.

The policies should demonstrate currency with technology (e.g., no mention of floppy drives), specificity to the environment (not off-the-shelf or AI generated), and be reviewed by the policy owner and approved by the appropriate role or committee at least annually.

Certifications

Some audits are specifically to provide certification that an organization's infrastructure meets a particular standard. Aligning and certified are not the same—alignment indicates the program is focused on and may meet all the requirements of a standard, whereas a certification is verification from a properly qualified assessor. For example, an ISO 27001 certification may only be granted by

---

[122] See the Policies and Procedures section.

organizations who have completed specific processes as mandated by ISO.

Bottom Line: Corporations do not absolve themselves of information security risk by engaging TPSPs. Indeed, the ability to ensure that the TPSPs have controls in place to protect information is sometimes more difficult than evaluating the same in-house. Independent audits help bridge this gap but may not be all that is needed for proper due diligence.

SMB Considerations: SMBs should also perform annual vendor reviews, to include requesting SOC2 audit reports. This provides a measure of assurance of controls as well as proof of due diligence should something go wrong.

Human Resource Considerations

Many in information security view staff as the weakest link in information security.[123] While some may consider that hyperbole, it's hard to argue that staff don't have any effect on information security. Thus, regardless of your position, ensuring security considerations when recruiting and hiring is essential.

Criminal Background Check

To protect your business, you need to understand the risks regarding people you add to your staff. Be it employee or consultant, you should conduct a criminal background check at minimum.

This is not to say that a candidate with a criminal history should be discounted. The criminal justice system is geared to reform. There are several superstars in information and cybersecurity who don't have the cleanest of records.[124] These highly skilled individuals have gone on to make significant contributions to the industry.

---

[123] https://www.isaca.org/resources/isaca-journal/issues/2019/volume-5/the-human-factor-in-information-security

[124] Kevin Mitnick may have been the most famous—https://en.wikipedia.org/wiki/Kevin_Mitnick

The organization needs to be able to completely understand the risks of bringing on staff. This requires complete honesty. If candidates have a blight on their record, rightfully or wrongly accused, it is best they be upfront about it. Many consider dishonesty an automatic deal-breaker. For information security where trust and ethics are foundational it is even more so.

Basic background checks are not expensive and there are many options for SMBs. The checks in the United States should include a Social Security Number trace, criminal record check, and sex offender check at minimum. Many background check organizations offer several options.[125]

Education and Employment Check

Unfortunately, misrepresenting education credentials happens. A LinkedIn check may not be enough, as there is no verification of anything posted on a LinkedIn user's profile. Education as well as employment and military service can be misrepresented. Again, honesty is a critical attribute of a quality employee.

Reduce the likelihood of the risk of hiring one misrepresenting credentials, and possibly putting the organization's information security at risk, by performing education and/or employment checks on candidates.

---

[125] Checker is one example—see https://checkr.com/pricing

Credit Check

In some instances, credit checks may be appropriate. This is usually the case in financial institutions such as banks and credit unions, where employees may have access to systems that handle financial transactions.

If a candidate does not have a good credit rating, it may be indicative of financial difficulties. They may be attempting to secure a position where they can embezzle funds to relieve their financial pressures.

Information Security Policy Acknowledgement

A new employee, contractor, or volunteer should not have access to information and systems before reading and agreeing to the company's policy or policies that govern information security. This practice is to reduce the risk of unknowingly causing an information security event or incident.

As noted earlier, the information security and acceptable use policies are two foundational documents that direct the protection of information and information systems use respectively. While not all policies that address information security elements (such as a secure coding policy which is limited in applicability to information technology) require all staff acknowledgement, those that are should be reviewed and acknowledged on hire and annually.

The review should also be documented. For small organizations, this can be as simple as noting on a spreadsheet. However, spreadsheets do not track action items and date stamps for audit purposes. Payroll and Human Resources systems such as Gusto[126] can serve as a current policy repository and an auditable method of recording reviews.

Sanctions for Violating Policies

Policies are like laws in that they have little effect if there are no consequences. Sanctions could include additional awareness, disciplinary action, or termination. The human resources department sets the parameters for policy violation response.

Training

Corporate-wide training is essential for reducing information security threats. As noted previously in the Governance section, staff behavior is often the weakest link in the information security risk management chain. The best way to mitigate this risk is through proper training. Additionally, the training should be interesting and promote engagement, not an exercise in compliance.

Basic Information Security Awareness Training

At minimum, annual basic information security awareness training provides recurring support of simple principles, such as identifying phishing emails. It usually is a baseline

---

[126] https://gusto.com/

requirement for regulations and standards. The training does not have to be overly intense—a typical annual refresher may take thirty minutes. Additionally, there are many tools available for managing training programs referred to as learning management systems (LMS).[127] Leveraging an LMS means it is not necessary to develop curricula in house.

One advantage of a LMS is the ability to track a user's training record. Some combine courses with phishing campaign results to provide a scoring based on risk. Increasingly the LMS field is incorporating risk information related to email addresses from sites that track when addresses are compromised[128] to enhance the user's risk score.

Targeted Training

Some roles require additional information security training. For example, it's common for developers to undergo annual OWASP Top Ten[129] training. Other targeted training could include information technology personnel and executives

---

[127] In addition to KnowBe4 (www.knowbe4.com) mentioned before, Wizer is a security awareness training platform focused on small businesses—see https://www.wizer-training.com/.

[128] For example, https://haveibeenpwned.com/.

[129] https://owasp.org/www-project-top-ten/

(who are often targets of business email compromise[130] scams).

Compliance Training

Often training centered on a particular regulation is necessary for both security and compliance. Many commercial training platforms will offer training to meet these needs, though likely at an additional cost. However, cost should be just one factor in determining the need; the primary driver should be the risk lack of training poses to the organization.

If such modules are not available with your training system, it may be possible to upload company materials for the organization's use. Some offer sharable content object reference mode (SCORM) compatibility, ensuring a course the organization creates may be used seamlessly on different SCORM-capable LMS. But even if your LMS is not SCORM capable, or creating a training module that is SCORM compatible is not a preferred option, slide deck presentations (e.g., PowerPoints) are usually accepted.

Providing regulatory compliance training (e.g., PCI, HIPAA) within an LMS is advantageous versus simply sending a slide deck to staff. The LMS can track when the course was taken

---

[130] https://www.fbi.gov/scams-and-safety/common-scams-and-crimes/business-email-compromise

and, if a quiz is offered, whether they passed the training. This then becomes a part of the staff training record.

Phishing Training

Phishing is a scam by which an internet user is duped by a deceptive email message into revealing personal or confidential information which the scammer can use illicitly[131]. Often the criminals are after credentials to corporate systems, which is why implementing MFA[132] is a very effective control in combating phishing scams and other fraud.

Organizations should periodically test staff phishing responsiveness by sending specially crafted test emails designed to see if an employee or other staff clicks on a phishing link. Doing so should not result in punitive action[133] but should instead be an indicator to provide additional education and awareness training. Phishing exercises can contain relevant content (e.g., reportedly providing COVID-19 information during the 2020—2021

---

[131] https://www.merriam-webster.com/dictionary/phishing

[132] https://www.nist.gov/itl/applied-cybersecurity/tig/back-basics-multi-factor-authentication

[133] I have heard of organizations where the first click resulted in a reduction or elimination of bonus pay. Such draconian responses result in paralyzed fear and a culture of distrust of the (v)CISO and is therefore not recommended.

pandemic[134]) and can vary in difficulty level depending on platform used to run the phishing campaigns.

Other Training

Other training could be external such as AICPA-sponsored events for finance personnel or internally provided by the (v)CISO or other security, technology, or risk management personnel. Regardless of the delivery system, the training should be meaningful and interesting, and not just a "check the box" exercise.

Bottom Line: There will never be a technology or "black box" solution that eliminates human behavior risk. However, proper training can substantially reduce that risk.

SMB Considerations: SMBs have the same training needs as large organizations, since they face many of the same threats. Fortunately, a quality information security awareness program is not costly to implement and maintain.

---

[134] The COVID-19 pandemic will not be forgotten, nor should its lessons. The COVID Tracking Project has been a constant source of COVID-19 information—see https://covidtracking.com/data.

Ethics

Information security is built on a fundamental foundation of trust. When this trust is not at the forefront of every information security action, gaps and risks in the security program can result.

SMBs need to have assurance that the resources they utilize for managing and enhancing their information security program and posture practice the highest level of professional ethics. The certification body ISC2 requires all certified security professionals to adhere to a code of ethics that "is a collection of requirements that apply to how you act, interact with others (including employers) and make decisions as an information security professional. The code is designed to 'give assured reliance on the character, ability, strength, or truth of a fellow ISC2 member, and it provides a high level of confidence when dealing with a peer member.'"[135]

As an example, in a LinkedIn post a consultant asked colleagues for a copy of a standard that cost several hundred dollars. I decided to reach out to the poster privately to note this action was not ethical, but by then

---

[135] https://resources.infosecinstitute.com/certification/the-isc2-code-of-ethics-a-binding-requirement-for-certification/

they had already deleted the post. I can only assume someone else pointed out the ethical violation.

If a consultant can justify stealing—and procuring an unauthorized copy of a standard is stealing—then how will they serve your business? Will they cut corners because they don't feel that they must do the work? Will they place your organization in a precarious position by "borrowing" your data without your knowledge?

While this may seem farfetched, the truth is that information security is a highly competitive industry. There are, unfortunately, those who believe "rules for thee and not for me."

SMBs rely on their vCISOs to have the highest of ethics. Your vCISO should have earned at least one respected and well-known certification such as ISC2's Certified Information Systems Security Professional (CISSP) that includes a code of ethics as part of the certification as noted above. Known violations may be reported and certification revoked.

All service providers should either keep their client list confidential, explicitly state in agreements otherwise, or obtain permission from clients for one-off requests (such as reference checks). While it may seem like unnecessary friction in the RFP and RFI response processes, many firms will not release client contact information for such purposes without an executed MNDA.

Bottom Line: There are no shortcuts. Those that try to subvert processes by bending ethics have their own, not their clients', best interests at heart.

SMB Considerations: SMBs require their service providers to uphold the highest ethical standards. Moreso than large businesses, SMBs often have very little budgetary room for dealing with mistakes from cutting corners. The adage "measure twice and cut once" applies here.

Business Continuity, Disaster Recovery, and Incident Response

Many use the terms business continuity, disaster recovery, and incident response interchangeably. However, each of these components of the resilience triad[136] serve different purposes, and understanding the role and interactions of each is essential to maintaining a robust information security program.

Business Continuity

"Business continuity is an organization's ability to maintain essential functions during and after a disaster has occurred. Business continuity planning establishes risk management processes and procedures that aim to prevent interruptions to mission-critical services and reestablish full function to the organization as quickly and smoothly as possible."[137]

Business continuity can be argued to be the most important component of the resilience triad; disaster recovery and incident response support it. However, of the three it is

---

[136] As far as I know, and as of the writing of this edition, I am not aware of a descriptor to encompass all three, yet I needed one for purposes of explaining this section. Thus, I made up this term. Perhaps it will stick and serve as a small cybersecurity nomenclature contribution.

[137] https://www.techtarget.com/searchdisasterrecovery/definition/business-continuity

often the most overlooked, because it is assumed, incorrectly, that the other components adequately ensure continuity of operations. They are important, sure, but are not complete.

For example, unique to the business continuity plan (BCP) is the inclusion of business impact analysis (or analyses, as often individual departments also perform their own). Shortened as BIA, they provide guidance for the disaster recovery (DR) actions. In the most simplistic example, BIAs direct information technology on service restoration prioritization based on business needs.

There are many methods for completing a BIA, and the mechanics are not as important as the goals of service restoration with minimal adverse revenue and reputational impact. While components may vary, BIAs should at minimum contain the following:

- Identification of critical systems (e.g. online banking portal);
- Identification of critical information;
- Identification of stewards of the business process;
- Identification of dependent processes; and
- Inclusion of recovery time objectives (RTO), the acceptable downtime for a function before the business impact becomes significant, and Recovery Point Objective (RPO), the acceptable amount of data that can be lost in restoration processes before the business impact becomes significant.

The BCP therefore should incorporate the BIAs to create a restorative process that prioritizes systems based on criticality to business operations. Too often information technology is left to make this decision in a vacuum. Information technology supports and needs to understand the business, but should never make risk management decisions for the business.

The BCP should be tested periodically. Business continuity plan tabletop exercises (BCP-TTX) are simulations of one or more events that may disrupt business operations, such as a tornado striking a primary data center. The BCP-TTX is designed to test incident response and expose gaps in that response.

The BCP-TTX is often one of the most overlooked elements of a holistic information security program. It involves time and effort from high-ranking managers and executives across the organization and is easily dismissed due to "scheduling conflicts." Unfortunately, not conducting at least one annual BCP-TTX robs the organization of the benefits of the exercise, while also not meeting the requirements of some frameworks and regulations.

An example of the importance of the BCP-TTX process was demonstrated in 2018 when my firm led its clients through a pandemic simulation. There wasn't much excitement in the exercises, as the general opinion was the chance of a pandemic that could impact business operations was nil. Previous pandemics such as H1N1 and the Avian Flu did not significantly disrupt business operations; in fact, often it seemed that the most impact they had for businesses was for regulators to require a pandemic response plan. The COVID-19 pandemic, of course, proved these assumptions wrong.

Tabletop exercises, as noted before, can expose gaps in information security programs. The COVID-19 pandemic resulted in a large percentage of the global workforce suddenly shifting to working from home. It is indeed fortunate that the infrastructure existed to support video conferencing on such a grand scale. This likely wouldn't have been possible as recently as ten years earlier, given the technology and infrastructure of the day. Yet the sudden shift resulted in serious, immediate information security concerns such as how to enable and confirm secure network access, and even paper document disposal.

The clients who participated in the pandemic tabletop exercise were much better prepared for the sudden shift to working from home. In fact, each of those clients now see BCP-TTX in a different light. It may be a stretch to say that they eagerly anticipate the annual BCP-TTX, but they certainly have learned the business value of it.

Disaster Recovery

The next component of the resilience triad is disaster recovery, or DR, the process by which information technology systems are restored. It is often, and incorrectly, confused with business continuity (BC). DR supports BC by ensuring that systems and services are restored as directed by the business continuity plan (BCP).

An unfortunate yet all-too-common situation is when an organization does not have a BCP. In that instance, information technology professionals are left to decide what systems are brought online first. Some of this is necessary due to the dependencies of systems; for example, it makes little sense to restore a public web site hosted internally before restoring internet connectivity back.

However, for any business to place the responsibility of risk management on the shoulders of information technology is irresponsible, especially at a time of urgency. Information technology will likely make decisions based mostly on system dependency. Moreover, many in information technology are not risk management professionals and should not be expected to take on that task, and associated liability. Unless risk management is in their job description and expectations are clearly communicated and managed, this can present a significant gap for any organization, especially SMBs.

As noted, DR actions should be informed by and follow the BCP. Other items of a well-constructed DR plan (DRP) include current versions of:

- System administrator list,
- Contact list,
- Network documentation,

- Wiring documentation, and
- System documentation.

There is an obvious commonality of these components. A DRP, possibly more so than the other resilience triad elements, is subject to changes over time by the dynamic nature of information technology. Therefore, the DRP should be reviewed regularly. Quarterly would not be too aggressive if the environment is rapidly changing.

DRPs should also be tested periodically. While this could be as a component of a BCP-TTX, since the latter is more about business processes and less about technology, DRPs are often better served with independent and focused testing. Additionally, as DRPs sometimes involve external entities (e.g., a bank coordinating with a core provider), scheduling can become challenging with more parties involved.

Incident Response Plan

An incident response plan, or IRP, directs how the business will manage an incident as it unfolds. Incidents rarely affect internal operations and personnel only; clients, customers, citizens, and other partners and consumers of your business are critical stakeholders during an incident.

Often external stakeholders are stressed when your organization is in the middle of an incident. They may rely on your business to serve their customers, or to maintain business operations, or to manage personal lives.

Companies have been severely impacted, if not ruined, by not managing an incident effectively.[138]

To craft an effective IRP, a definition of "incident" needs to be adopted and understood by all stakeholders, internal and external. In information security, an incident is often described as one or more series of information security events that has resulted in a situation where there is significant likelihood that information or systems could be compromised. An information security event is an observed action that may lead to an incident. A basic example is where logging systems record several events of an external entity attempting a brute force attack[139] on a system, and traffic analysis shows connectivity between the system and an external, unknown entity. Collectively, these events point to a high likelihood that the system was compromised.

One component of an IRP that should never be overlooked is not technical. The business should have one and only one source of communication during an incident. Conflicting messages may confuse customers during a period when their confidence needs management and restoration.

---

[138] Code Spaces was one famous example—see https://www.csoonline.com/article/2365062/code-spaces-forced-to-close-its-doors-after-security-incident.html.

[139] A trial-and-error process of attempting to log into accounts without authorization by trying different passwords.

Additionally, incorrect information could have contractual and legal implications. Also, incident details may aid the intruder, providing information to pivot attack methods. All communication must be vetted and approved by executive management and cleared through legal.

Other components are trigger points, or situations that direct a specific decision. One example is a ransomware infection and when or if to pay the ransom. A trigger point may be if systems will be down more than the defined RTO, resulting in potential losses exceeding the organization's risk tolerance. In the heat of the moment a business should not be making such decisions without some prior planning. Often the cyber liability insurance carrier will have a say in this as well.

Organizations should craft their IRP with different scenarios (e.g., ransomware, pandemic). The choice of playbooks should be based on the realistic risks the organization may face. For example, a business located in Kansas may want to have a plan drafted to direct response actions in case of a tornado but likely won't need to consider one for a hurricane.

Remember, like business continuity, incident response is not solely an information technology issue. Don't treat it as such.

Bottom Line: Understanding the resilience triad and paying each its proper due diligence on an annual basis is necessary; it will govern what happens when an actual disruptive incident occurs. No business should attempt to make decisions on the fly when timely restoration is critical for business survival.

SMB Considerations: SMBs may be more susceptible to a business disruption, even to the point of threatening business survival. From that viewpoint, proper resilience triad management may be more valuable, comparatively, for SMBs.

Strategic Planning

Earlier I touched on the importance of reporting on the information security program to the board of directors by relating to the corporate strategic plan.[140] Doing so may be easier and clearer to the board if there is an information security strategic plan in place.

An information security strategic plan provides a roadmap and structure for determining and managing an organization's information security needs and goals. It is a high-level document and should not contain tactical items. At minimum, it should align with the organization's corporate strategic plan. Ideally, specific information security strategic goals should map to corporate strategic goals.

The plan also demonstrates to third-party auditors that the information security program is in a constant state of evaluation and improvement. Additionally, some regulations and standards (e.g., PCI DSS, ISO 27001, and HIPAA) require that an information security strategic plan or equivalent be in place.

There are many approaches to developing such a plan, but all generally look at one or more future points, such as one,

---

[140] See the Governance section.

three, and five years out.[141] The further out, the less accurate the plan will be. Particularly in information security, threats, technologies, and business needs are not static. Therefore, an information security strategic plan should be updated every year to account for environmental changes.

The information security strategic plan should consider the information security risk assessment as discussed earlier.[142] A rule of thumb would be that strategic risk mediation has an inverse relationship to time. In other words, the highest strategic risks should be addressed in year one of a one-three-five plan, whereas low strategic risks don't need to be addressed until year five (or later).

While the development of an information security strategic plan may seem to be straightforward, mistakes are often made, leading to inadequate processes. The first is focusing on information technology only. I discussed in the beginning of this book[143] that information technology security is a subset of information security. Therefore, focusing strategic planning on only information technology

---

[141] Often referred to as a "one-three-five plan."

[142] See the Risk and Risk Assessment section.

[143] See the Information Technology Security and Information Security section.

security will result in missing strategic concerns for holistic information security risk management.

Another is not mapping to the corporate strategic plan. Information security supports the business, not the other way around. All business unit operations, including information security, need to map to and support the central strategic plan. This can become a real problem if there is no corporate strategic plan. Not only is that an information security risk, it puts the responsibility of determining strategic goals on the information security team.

Finally, some organizations may find themselves in a perpetually reactive state. Reactive organizations who only focus on addressing gaps never have an opportunity to look forward. This is a difficult hill to climb and may require increased resources up front to fill gaps while planning for the future.

Bottom Line: Without a forward-looking plan, information security actions may be inefficient and incomplete. As with all business planning, information security needs to be strategically and carefully addressed, looking at both short and long term, in accordance with corporate goals.

SMB Considerations: Some SMBs may not require a formal corporate strategic plan but they certainly benefit from a documented process. Businesses of all sizes need to set goals; otherwise, they are aimless in their directions. The information security strategic plan is just as important for SMBs as for larger organizations and should not be siloed.

GRC Systems

Governance, risk, and compliance (GRC) systems[144] help to manage risk-based information security programs. For small organizations, tracking elements of the information security program is often accomplished successfully via spreadsheets. This is a siloed approach that does not scale well, however.

Information security as risk management encompasses many elements of the organization. This means that relationships among these elements—risks to assets, for example—are often as important as the elements themselves. While spreadsheets are adequate for tracking segmented items, they do not work well when describing and managing relationships.

That's why GRC systems were developed. A GRC system can aid in managing the information security program holistically by leveraging relationships such as:

- Audits to assets,
- Information to risks, and
- Framework compliance to controls.

While a GRC system setup can be quite lengthy and involved, a well-tuned GRC spays off in dividends. Gone are

---

[144] Also referred to as integrated risk management (IRM) systems.

the days when hours were spent comparing spreadsheets or assessing risks to systems that may no longer exist (or worse, not assessing risks to systems that have been installed but have not been communicated to the information security department).

When choosing a GRC system, the business should first decide what business issues they are trying to solve initially, then decide if the other aspects of a GRC system can add value above the cost to the organization.

Bottom Line: GRC systems help to effectively manage a security program. Spreadsheets are siloed and do not scale because components of information security risk and compliance are related, not siloed.

SMB Considerations: GRC systems can help SMBs organize and manage their information security, risk management, and governance programs. There are solutions available that scale to SMB budgets.

Virtual CISO

An information security program is not complete without appropriate, experienced leadership. In the past, information security was often a component of information technology. However, more organizations, including SMBs, recognize that the information security role is larger than firewall management and represents a critical business function.

Larger organizations address this by employing a full-time CISO, often with a team of direct reports. However, SMBs often don't have the resources to hire a full-time CISO, nor do they require one. The virtual CISO, or vCISO, emerged as a solution to this gap. A vCISO is a consultant who assists organizations in creating, managing, and improving their information security program.

SMBs typically engage a vCISO in one of three ways:

- Project-based with a defined scope and deliverables, such as a risk assessment or SOC2 audit support,
- Retainer (prepaid) to serve as a trusted resource on demand, or
- Ongoing as a resource for continuous program support for a set number of hours per month (usually fifteen to forty).

If the need is for a resource more than forty hours a month, often hiring a full-time CISO is more cost-effective.

A vCISO is not necessary for SMBs to create and manage an effective information security program. Resources, such as this publication, can help savvy SMB owners and executives create and manage their program. However, if information security risk is not a core competency for a business, engaging a vCISO may be more efficient and effective to manage information security risk in the long run.

Qualifications

An effective vCISO is an information security executive with years of experience as the full-time senior information security executive at a midsized or larger company. This provides the seasoned vCISO with the practical real-world experience to serve SMBs. The vCISO is not selling products or tactical services, rather their product, and their value, is their experience.

The virtual CISO market has exploded over the past several years, and for good reason—it helps to solve a problem. However, with this growth has come dilution. Some providers are offering vCISO services with unqualified resources, which in turn presents a risk to the business. To minimize the chance of engaging such, consider the following actions to take when vetting virtual CISOs:

- Understand and verify the resource's experience (whether it be a single vCISO or a service provider),

- Ensure they are certified by a respected and well-known organization that includes the requirement of adhering to a code of ethics,[145] and
- Request proof of sufficient errors and omissions (E&O) insurance which "is a form of liability insurance that covers your business against claims of mistakes in professional services, such as services that are late, never delivered or inaccurate."[146]

A vCISO, by definition, works remotely, but may be available for on-site meetings such as in the support of an audit. Additionally, a vCISO usually works with several clients at a time and is often not available at a moment's notice.

Virtual CISOs should not be leveraged as a resource for immediate incident response (boots on the ground) when an incident occurs,[147] and instead vCISOs support incident management preparation with planning, process development, and post-recovery to include root cause analysis and process improvements. Additionally, a vCISO

---

[145] See the Ethics section.

[146] https://www.forbes.com/advisor/business-insurance/errors-and-omissions/

[147] Many virtual CISO providers partner with firms that specialize in incident response to maintain separation of duties and provide the specialty skillset needed.

does not perform forensics or install and configure information technology security equipment.

Cost

A vCISO may be an appropriate solution given budget constraints, but a highly experienced vCISO (one with years of full-time experience as a CISO and/or significant vCISO experience) will charge a commensurate fee.

The hourly rates for virtual CISOs can vary widely. As a rough guide, you can expect to pay anywhere from $200 to $500 per hour for virtual CISO services, depending on a range of factors, including:

- Experience: More experienced vCISOs are likely to command higher hourly rates, as they bring a wealth of expertise and knowledge to the table.
- Organization size and complexity: Larger and more complex organizations may require more extensive cybersecurity support and will therefore be expected to pay higher hourly rates for vCISO services.
- Scope of the project: The scope of the project will have a major impact on the hourly rate charged by vCISOs. For example, a full-scale cybersecurity audit is likely to cost more than a simple security assessment.
- Level of expertise required: Projects that require specialized knowledge or advanced technical skills

may command higher rates, particularly for regulated industries such as finance and healthcare.

Engagement

Engaging a vCISO often begins with a request for proposal (RFP) or request for information (RFI).[148] A properly constructed and effective RFP or RFI should convey what the business needs. This should include immediate and long-term requirements.

A discovery call, usually a half-hour in duration, helps to determine the scope of the engagement. During this, the virtual CISO firm will ask questions about the request or, if there is no formal RFP or RFI, will attempt to gain details to properly scope a proposal.

Some firms offer services on an hourly basis while others engage a flat rate. Generally, the latter is the industry norm, but prepaid engagements such as retainers may involve an hourly rate.

A typical initial term for a vCISO engagement is one year. The first three months usually involve learning the "as-is" of the environment. A flat rate normalizes expenses and reduces the overheads incurred by hourly billing. Flat rate is generally the most cost-effective billing method.

---

[148] See the Create and Evaluate the RFI or RFP section.

Conflict of Interest

As providing vCISO services has become both popular and profitable, some service providers, as a revenue growth strategy, include virtual CISO services in their offerings not primarily to increase service to clients, but rather as an inside sales approach. To illustrate this, consider a typical vCISO engagement.

When conducting a gap analysis against an applicable framework, the vCISO, if not truly independent from other books of business of their organization, may uncover gaps that the service provider offers solutions to remediate. If you've realized this is a potential separation of duties issue, crossing second and first line,[149] you are correct.

Additionally, since the service provider may be more focused on upselling their services (if that is their main revenue generator), their vCISO may downplay other potential gaps that the service provider cannot directly remediate. Engaging a potentially biased vCISO, therefore, can result in bad advice and unknown gaps, in addition to potential costs for technical solutions that may or may not be needed.

This is not to say that all service providers operate in this way. Those who do not are the ones who understand the

---

[149] The Three Line of Defense Model is discussed in more detail in the Reporting Structure section.

value of unbiased consultants. A simple check would be to ask for verification that the organization does not require the use of their services to resolve gaps identified by their vCISO.

Alternately, the business can decide to accept the bias, noting that a combined solution offers several advantages, to include a single point of contact and a lower likelihood of system incompatibility. Service provider transparency will help the business make this decision.

Success

Your vCISO is your business partner. Honesty, transparency, and participation are necessary in all phases of the engagement. The vCISO is positioned to improve the business' information security posture through proactive means; this can only occur if the vCISO is seen as part of a team.

This is not an audit engagement. The work they produce is for the client only and not generally intended to be used as a third-party attestation. The vCISO will help prepare your organization for an audit but will not perform one. This maintains independence, plus these are two different skill sets.

An effective vCISO will tell the client what they need to hear, not what they want to hear. Unfortunately, some may downplay or withhold risk information, believing that good news shows progress and therefore may lead to a longer engagement.

Ultimately, whether a vCISO is the best option for your organization's information security operations is a business decision. Research the field and ask questions. Most who practice this trade have a genuine desire to help SMBs and will gladly point you in the right direction for your business.

Bottom Line: Vetting and understanding the virtual CISO service your business is contracting for, as well as maintaining transparency and participation, is critical for success.

SMB Considerations: SMBs often have the need but not the resources for a full-time CISO. That's where a properly vetted and highly experienced virtual CISO adds value. This is not an area to go cheaply; in the long run, the cost of fixing mistakes or breach responses that could have been avoided (or both) is usually much more expensive.

Printed in the USA
CPSIA information can be obtained
at www.ICGtesting.com
LVHW021555080824
787585LV00014B/381

9 781733 066860